U.S. Department of Transportation
Research and Innovative Technology Administration

Implementation and Evaluation of Weather Responsive Traffic Estimation and Prediction System

Contract No: DTFH61-06-D-00005
Task No: T-10-005

Final Report — June 18, 2012
FHWA-JPO-12-055

Produced by Northwestern University for SAIC
ITS Joint Program Office
Research and Innovative Technology Administration
U.S. Department of Transportation

Notice

This document is disseminated under the sponsorship of the Department of Transportation in the interest of information exchange. The US Government assumes no liability for its contents or use thereof.

Technical Report Documentation Page

1. Report No.	2. Government Accession No.	3. Recipient's Catalog No.
FHWA-JPO-12-055		

4. Title and Subtitle	5. Report Date
Implementation and Evaluation of Weather Responsive Traffic Estimation and Prediction System	June 2012
	6. Performing Organization Code

7. Author(s)	8. Performing Organization Report No.
Hani S. Mahmassani, Jiwon Kim, Tian Hou, Ali Zockaie, Meead Saberi, Lan Jiang, Ömer Verbas, Sihan Cheng, Ying Chen and Robert Haas	

9. Performing Organization Name And Address	10. Work Unit No. (TRAIS)
SAIC M/S E-12-3 8301 Greensboro Drive McLean, VA 22102 Northwestern University The Transportation Center 600 Foster Street Evanston, IL 60208-4055	
	11. Contract or Grant No.
	DTFH61-06-00006, Task T-10-005

12. Sponsoring Agency Name and Address	13. Type of Report and Period Covered
U.S. Department of Transportation Federal Highway Administration Office of Operations 1200 New Jersey Avenue, SE Washington, DC 20590	Final Report August 2010 – June 2012
	14. Sponsoring Agency Code

15. Supplementary Notes
Roemer Alfelor - Government Task Manager

16. Abstract
The objective of the project is to develop a framework and procedures for implementing and evaluating weather-responsive traffic management (WRTM) strategies using Traffic Estimation and Prediction System (TrEPS) methodologies. In a previous FHWA-funded project, a methodology was developed and tested for incorporating weather impacts in off-line TrEPS. This capability is now included in on-line TrEPS, which interacts with multiple sources of local real-time information, to provide operators with predicted traffic states under the current and future weather conditions. The main goal is to support the decision making process for addressing the disruptive effect of inclement weather on the traffic system. These tools were developed, applied, calibrated and tested in three different locations across the US. The networks selected include Salt Lake City, UT, New York's Long Island Expressway Area and Chicago, IL. In addition to building the networks and developing the various demand and supply inputs, calibration of the traffic models to local weather conditions has provided a rich database of calibrated weather-sensitive traffic models that could be applied in any region. The methodology is configured to run with real-time traffic data (e.g. from loop detectors or RTMS sensors). The methodology was used to evaluate the effectiveness of different weather-related traffic management strategies in specific areas. The project has successfully demonstrated the potential of a TrEPS installation that is "always on", up to date and adaptively calibrated to address a range of weather conditions and scenarios, retrievable on demand by local operating agencies through a scenario manager.

17. Key Words	18. Distribution Statement
Weather-Responsive Traffic Management (WRTM), Traffic Estimation and Prediction, Weather and Traffic Analysis, WRTM Strategies, Dynamic Traffic Assignment	No restrictions.

19. Security Classif. (of this report)	20. Security Classif. (of this page)	21. No. of Pages	22. Price
Unclassified	Unclassified	182	N/A

Form DOT F 1700.7 (8-72)   Reproduction of completed page authorized

Acknowledgment

The study team would like to acknowledge the many individuals who have enthusiastically supported this study with their time, effort, and comments, and substantially contributed to its success. We are especially grateful to Roemer Alfelor and David Yang of the US Dept. of Transportation for their leadership in defining and guiding the study, and facilitating many contacts with local and state agencies. Robert Haas of SAIC, Inc. provided helpful comments and support throughout the study. Brian Park at the University of Virginia was a most helpful advisor especially in terms of evaluating the suitability of the Northern Virginia area as a potential test deployment site.

From the Utah Department of Transportation (UDOT), we would like to particularly thank Robert Clayton, UDOT Traffic Operations Center Director, for supporting this project and Glen Blackwelder, Traffic Operations Engineer of UDOT, and several other staff including:

- John Haigwood, Freeway Operations Engineer
- Leigh Sturges, Weather Operations Manager
- Tam Southwick, Traffic Mobility Engineer
- Rudy Zamora, IT Manager
- Tam Southwick, Traffic Mobility Engineer
- Audra Yocom, Utah PEMS Operator

From the New York State Department of Transportation (NYSDOT), we would like to particularly acknowledge Emilio Sosa, Director of the NYSDOT Region 10 (Long Island), Todd B. Westhuis, Director of the Office of Traffic Safety and Mobility, NYSDOT Operations Division, and John Bassett, Director of the System Optimization Bureau, NYSDOT.

From the New York City Department of Transportation (NYCDOT), we would like to gratefully thank Jeffrey Malamy, Director of Special Projects, Office of the Commissioner and Mohamad Talas, Deputy Director of the System Engineering, NYCDOT.

From the City of Chicago Department of Transportation (CDOT), we would like to particularly acknowledge David Zavattero, Deputy Director of CDOT and Abraham Emmanuel, Principal Systems Programmer of CDOT.

We would also like to thank Timothy P. Harpst, Transportation Director of the Salt Lake City Transportation Division and Kenneth T. Louie, Traffic Engineer, City of Irvine.

The authors remain solely responsible for all work, findings, conclusions and recommendations presented in this report.

Table of Contents

Executive Summary 8
1. Introduction 11
   1.1 Background 11
   1.2 Integration of Weather in TrEPS 12
   1.3 Weather Responsive Traffic Management Strategies 15
      1.3.1 Advisory Information 15
      1.3.2 Control Strategies 15
   1.4 Conceptual Framework for On-line Implementation 17
      1.4.1 Implementation of WRTM strategies using TrEPS models 17
      1.4.2 Evaluation of WRTM strategies using TrEPS models 18
   1.5 Project Approach 21
   1.6 Structure of Final Report 22
2. Study Networks 24
   2.1 Site Selection Procedure 24
      2.1.1 List of Candidate Sites 24
      2.1.2 Criteria for Network Selection 25
      2.1.3 Selection Results 33
   2.2 Network Preparation 36
   2.3 Data Collection 40
      2.3.1 Weather Data 40
      2.3.2 Traffic Data 44
3. Calibration and Validation of Weather-sensitive TrEPS Model 48
   3.1 Supply-side Parameter Calibration 48
      3.1.1 Calibration of Traffic Flow Model Parameters 48
      3.1.2 Calibration of Weather Adjustment Factor (WAF) 55
   3.2 Demand-side Parameter Calibration 60
      3.2.1 Estimating Base Case OD Matrix 60
   3.3 Validation of Weather Sensitive DYNASMART-P 72
      3.3.1 Validation Procedure for Weather Specific Simulations 72
      3.3.2 Validation Results 73

4. Identification of Existing WRTM Strategies 76
4.1 Background 76
4.2 Identifying Existing WRTM Strategies Used By Selected Agencies 77
4.2.1 Survey Design 77
4.2.2 Survey Results 77
4.3 Recommendation of Candidate WRTM Strategies for This Study 80
4.3.1 Basis for Recommendation 80
4.3.2 Recommended WRTM Strategies for Each Network 81

5. Evaluation Approaches to Assess Benefits of WRTM 85
5.1 Purpose 85
5.2 Performance Measures 88
5.2.1 Network-level Measures 88
5.2.2 OD/Path-level Measures 91
5.2.3 Link-level Measures 96
5.2.4 Cross Section Measures 97

6. Implementation and Evaluation of Selected WRTM Strategies for Study Networks 99
6.1 Chicago Network 99
6.1.1 Coordination with Chicago DOT 99
6.1.2 Sub-network Preparation 100
6.1.3 Implementation and Evaluation of WRTM Strategies: (1) Demand Management 102
6.1.4 Implementation and Evaluation of WRTM Strategies: (2) Variable Speed Limit 106
6.1.5 Implementation and Evaluation of WRTM Strategies: (3) Optional Detour VMS 113
6.2 Long Island Network 117
6.2.1 Coordination with New York State DOT 117
6.2.2 Sub-network Preparation 117
6.2.3 Implementation and Evaluation of WRTM Strategies: (1) Incident Management 121
6.3 Salt Lake City Network 135
6.3.1 Coordination with Utah DOT 135
6.3.2 Sub-network Preparation 135
6.3.3 Implementation and Evaluation of WRTM Strategies: (1) On-line Implementation 138
6.3.4 Implementation and Evaluation of WRTM Strategies: (2) Demand Management 146

7. Conclusion 151
7.1 Summary and Accomplishments 151

	7.1.1	Summary 151
	7.1.2	Key Accomplishments of the Study 153
7.2		Lessons Learned and Next Steps.. 154
	7.2.1	Lessons Learned 154
	7.2.2	Next Steps 157

8. References.. 160
Appendix A: Candidate Networks and Adjacent ASOS Stations 163
Appendix B: Distribution of Weather Stations (ASOS Stations vs. Clarus System(ESS)) 168
Appendix C: Calibration Results for Traffic Flow Model and WAF 171

List of Tables

Table 2-1. Characteristics of Candidate Networks (Traffic and Weather Data) 27
Table 2-2. Characteristics of Candidate Networks (Geographic Scope and Weather Pattern) 28
Table 2-3. Network Selection Criteria and Evaluation Results 34
Table 2-4. Airports with ASOS Stations and Available Time Periods for Data (Long Island, NY) 41
Table 2-5. Airports with ASOS Stations and Available Time Periods for Data (Chicago) 42
Table 2-6. Airports with ASOS Stations and Available Time Periods for Data (SLC) 43
Table 2-7. Airports with ASOS Stations and Available Time Periods for Data (Irvine) 43
Table 3-1. Weather categorization for different networks 51
Table 3-2. Supply Side Properties related with Weather Impact in DYNASMART 55
Table 3-3. Calibration results of WAF 59
Table 3-4. RMSE Values for the Long Island Network 64
Table 3-5. RMSE Values for the Chicago Network 67
Table 3-6. RMSE Values for the Salt Lake City Network 70
Table 3-7. RMSE Values for the selected snow scenario 73
Table 4-1. Six Categories of WRTM Strategies 77
Table 4-2. Survey Response from Agencies 79
Table 4-3. WRTM Strategy Recommendation for the Chicago network 82
Table 4-4. WRTM Strategy Recommendation for the Salt Lake City network 83
Table 4-5. WRTM Strategy Recommendation for the New York network 84
Table 6-1. Comparing Network Characteristics for Original and Extracted Networks of Chicago 101
Table 6-2. Constructing Variable Speed Limit Strategy Scenarios 108
Table 6-3. Selected Scenarios for Detailed Evaluation 111
Table 6-4. Selected Scenarios for Detailed Evaluation 116
Table 6-5. Comparing Network Characteristics for Original and Extracted Networks of Long Island 119
Table 6-6. Incident Scenario based on Historical Data 124
Table 6-7. Optional Detour VMS Scenarios 124
Table 6-8. Selected Scenarios for Detailed Evaluation 128
Table 6-9. Descriptive Statistics for Path Travel Time 132
Table 6-10. Comparing Network Characteristics for Original and Extracted Networks of SLC 137

List of Figures

Figure 1-1. Incorporating Weather in DTA Model (Source: Mahmassani et al., 2009) 13
Figure 1-2. DYNASMART-X, Real-time Traffic Estimation and Prediction System 14
Figure 1-3. Framework for Implementing WRTM strategies using TrEPS models 19
Figure 1-4. Framework for Evaluating WRTM strategies using TrEPS models 20
Figure 1-5. Key Steps for Implementation and Evaluation of Weather Responsive TrEPS System 22
Figure 2-1. Geographical Distribution of Candidate Sites (Source: National Climatic Data Center, NOAA) 25
Figure 2-2. Monthly Surface Data: Number of Days with Rain Prec. ≥ 0.1 inch 30
Figure 2-3. Monthly Surface Data: Number of Days with Rain Prec. ≥ 0.5 inch 30
Figure 2-4. Monthly Surface Data: Number of Days with Rain Prec. ≥ 1.0 inch 31
Figure 2-5. Monthly Surface Data: Number of Days with Snow Depth ≥ 0.1 inch 31
Figure 2-6. Monthly Surface Data: Total Monthly Rain Precipitation 32
Figure 2-7. Monthly Surface Data: Total Monthly Rain Precipitation 32
Figure 2-8. Network Rating 35
Figure 2-9. Flowchart for the Conversion from the Static to the Dynamic Network Model 37
Figure 2-10. Network Configuration and Description for New York Network (Long Island) 38
Figure 2-11. Network Configuration and Description for Chicago Network 38
Figure 2-12. Network Configuration and Description for Salt Lake City Network 39
Figure 2-13. Network Configuration and Description for Irvine Network 39
Figure 2-14. Long Island Study Area and Adjacent ASOS Stations 41
Figure 2-15. Chicago Study Area and Adjacent ASOS Stations 42
Figure 2-16. Salt Lake City Study Area and 43
Figure 2-17. Irvine Study Area and 43
Figure 2-18. Selected Detector Locations in Chicago 44
Figure 2-19. Selected Detector Locations in Salt Lake City 45
Figure 2-20. Selected Detector Locations in Irvine (Source: Google Map) 46
Figure 2-21. Locations for Selected Detectors and ASOS Station in Baltimore (Source: Google Map) 47
Figure 3-1. Type 1 modified Greenshields model (dual-regime model) 49
Figure 3-2. Type 2 modified Greenshields model (single-regime model) 50
Figure 3-3. Examples of raw traffic data and calibrated speed-density curves under different weather conditions for each network: Irvine (a,b), Salt Lake City (c,d), Chicago (e,f) and Baltimore (g,h). 54
Figure 3-4. Effect of the rain intensity on weather adjustment factors for: (a) maximum flow rate (q_{max}); (b) speed intercept (v_f); (c) breakpoint density (k_{bp}); and (d) free flow speed (u_f) 57
Figure 3-5. Effect of the snow intensity on weather adjustment factors for: (a) maximum flow rate (q_{max}); (b) speed intercept (v_f); (c) breakpoint density (k_{bp}); and (d) free flow speed (u_f) 58
Figure 3-6. Two Criteria in the Optimization Process 62
Figure 3-7. Observed and Simulated Counts on Selected Links (Long Island) 65

Figure 3-8. Temporal Distribution of SOV trips for the Long Island Network 66
Figure 3-9. Temporal Distribution of HOV trips for the Long Island Network 66
Figure 3-10. Observed and Simulated Counts on Selected Links (Chicago) 68
Figure 3-11. Temporal Distribution of trips for the Chicago Network 69
Figure 3-12 Observed and Simulated Counts on Selected Link (Salt Lake City) 70
Figure 3-13. Temporal Distribution SOV of trips for the Salt Lake City Network 71
Figure 3-14. Temporal Distribution of HOV trips for the Salt Lake City Network 71
Figure 3-15. Observed and Simulated Speeds on a Selected Link (Chicago) 74
Figure 3-16. Observed and Simulated Counts on a Selected Link (Chicago) 75
Figure 5-1. Evaluating WRTM strategies in Simulation and Real-world Environments 86
Figure 5-2. DYNASMART-X GUI; Prediction Results with and without Intervention 87
Figure 5-3. Visual Representation of Performance Measures for Chicago Network 89
Figure 5-4. Percentage of Lane-miles for Each Density Level (Scenario 1 vs. Scenario 2) 90
Figure 5-5. Percentage of Lane-miles for Each Speed Level (Scenario 1 vs. Scenario 2) 91
Figure 5-6. OD/Path Travel Time Distribution and Associated Measures 93
Figure 5-7. Comparing Travel Time Characteristics of Different Scenarios 94
Figure 5-8. Time-dependent Average Path Travel Time (top) and Average Travel Time Per Mile for Each Link along the Path (bottom) 95
Figure 5-9. Time-dependent Traffic Flow Performance Measures for Selected Link 96
Figure 5-10. Selected Cross Section on Long Island Network (Analysis for Westbound Traffic Flow) 97
Figure 5-11. Time-dependent Cross-section Throughput Measures 98
Figure 6-1. Map of the Extracted Network of Chicago 100
Figure 6-2. Temporal Distribution of Demand for Sub-network from 5:00 AM to 11:00 AM 102
Figure 6-3. Weather Scenario for Demand Management Strategy: 103
Figure 6-4. Accumulated Percentage of Out-Vehicle for Different Scenarios 105
Figure 6-5. Changes in Average Travel Time and Average Stop Time Relative to Benchmark 106
Figure 6-6. Weather Scenario for VSL Strategy: 107
Figure 6-7. Location of Variable Speed Limit Signs for Different Scenarios 108
Figure 6-8. Number of Vehicles Remaining in the Network at the End of Simulation 110
Figure 6-9. Total Travel Time in Hours for Different VSL Scenarios 110
Figure 6-10. Selected Cross Section for Measuring Traffic Throughput (Northbound) 112
Figure 6-11. Time-dependent Cross-section Throughput Measures (Cumulative Flows) 112
Figure 6-12. Location of Detour Signs for Different Scenarios 114
Figure 6-13. Number of Vehicles Remaining in the Network at the End of Simulation 115
Figure 6-14. Total Travel Time in Hours for Different Optional Detour VMS Scenarios 115
Figure 6-15. Time-dependent Cross-section Throughput Measures (Cumulative Flows) 116
Figure 6-16. Map of the Extracted Network of Long Island (Source: Google Map) 117
Figure 6-17. Network Extraction for Long Island 118
Figure 6-18. Temporal Distribution of 5hr-Demand for Sub-network (Single Occupancy Vehicles) 120
Figure 6-19. Temporal Distribution of 5hr-Demand for Sub-network (High Occupancy Vehicles) ..120

Figure 6-20. Snow Scenario based on Historical Data (extractedfrom2011-01-26 6:00AM – 12:00PM) 122
Figure 6-21. Incident Observations on I-495 WB (2011-01-26 06:00AM – 12:00PM) (Source: Google Map) 123
Figure 6-22. Incident Locations displayed in the DYNASMART network 123
Figure 6-23. Location of Detour Signs for Different Scenarios 125
Figure 6-24. Number of Vehicles Remaining in the Network at the End of Simulation 127
Figure 6-25. Total Travel Times Experienced by Vehicles at the End of Simulation 127
Figure 6-26. 511NY Traffic Information Service (VMS locations and associated messages) 129
Figure 6-27. Selected Two Locations for Measuring Path Travel Time 130
Figure 6-28. Time-dependent Average Travel Time for Selected Path 131
Figure 6-29. Average Link Travel Time Per Mile along Selected Path 131
Figure 6-30. Comparison of Travel Time Characteristics of Different Scenarios for Selected Path . 132
Figure 6-31. Selected Cross-Section for Measuring Traffic Throughput (Westbound) 133
Figure 6-32. Time-dependent Cross-section Throughput Measures (Cumulative Flows) 134
Figure 6-33. Time-dependent Cross-section Throughput Measures (Vehicle Counts/5min) 134
Figure 6-34. Map of Extracted Network of Salt Lake City 136
Figure 6-35. DYNASMART-X Connected to SLC Real-time Traffic Data Stream 139
Figure 6-36. Temporal Profile of 24-hr Demand for Salt Lake City Sub-network 139
Figure 6-37. Weather Scenario for On-line Implementation in DYNASMART-X 141
Figure 6-38. Locations of Variable Speed Limit Signs for Different Scenarios 142
Figure 6-39. Off-line Simulation Results for VSL Strategies 143
Figure 6-40. DYNASMART GUI during On-line Implementation (at 7:16 AM, April 6, 2012) 144
Figure 6-41. DYNASMART GUI during On-line Implementation (at 7:48 AM, April 6, 2012) 145
Figure 6-42. DYNASMART GUI during On-line Implementation (at 7:54 AM, April 6, 2012) 145
Figure 6-43. Weather Scenario for Demand Management Strategy: 147
Figure 6-44. Accumulated Percentage of Out-Vehicle for Different Scenarios 149
Figure 6-45. Changes in Average Travel Time and Average Stop Time Relative to Benchmark 150

Executive Summary

The disruptive effect of inclement weather on traffic results in considerable congestion and delay, due to reduced service capacity, diminished reliability of travel, and greater risk of accident involvement. To mitigate the impacts of adverse weather on highway travel, the Federal Highway Administration (FHWA) Road Weather Management Program (RWMP) has been involved in research, development and deployment of weather responsive traffic management (WRTM) strategies and tools. Dealing with adverse weather requires not only sensing of traffic conditions, but also the ability to forecast the weather in real-time for operational purposes. Incorporating weather effects and responsiveness in Traffic Estimation and Prediction System (TrEPS) models and software is an important capability for evaluation and deployment of Weather-Related Traffic Management (WRTM). The overall goal of the study is to implement and evaluate weather responsive traffic management strategies using Traffic Estimation and Prediction System (TrEPS) models. The TrEPS model selected for this study is DYNASMART-X, which is a real-time system that interacts continuously with loop detectors, roadside sensors and vehicle probes, providing real-time estimates of traffic conditions, network flow patterns and routing information. In this project, the weather-sensitive TrEPS model was applied, calibrated and tested in several major US cities. After conducting a systematic evaluation four final study sites were selected: Chicago, IL; Salt Lake City, UT; New York, NY; and Irvine, CA.

For the selected study networks, traffic and weather data were collected and supply- and demand-side parameters were calibrated. The study resulted in the development of a library of calibrated weather-sensitive traffic flow models and associated weather adjustment factors (WAF's) for different types of networks in different parts of the country; these provide starting point for rapid prototyping for any area in the US. In addition, the weather adjustment factors for a selected network (Chicago) were validated by simulating a specific weather scenario with and without using the WAFs. The test result confirmed that the use of WAFs successfully captures the weather effects on both link speeds and flows. Two types of WRTM strategy evaluations were performed: off-line simulation experiments using a weather-sensitive TrEPS planning tool to evaluate contemplated strategies, and on-line implementation of existing strategies to support the decision-making process under inclement weather conditions. On-line implementation for the Salt Lake City network achieved the intended demonstration of the prototype's capability to run with real-time input; however, the absence of significant weather events during the test period, due largely to an unseasonably warm winter, precluded extensive in-situ testing, which remains a topic for future tasks.

One of the important study findings is that the primary application of the TrEPS capability lies in the short-term operational planning and preparedness for inclement weather predicted to occur in the next 12 to 48 hours. As such, for each of the networks, the focus shifted on maintaining a calibrated on-call TrEPS models for the extracted subset of the network of interest, as this was the primary interest of the implementing agencies. An essential capability to enable such use is that of the scenario manager prototyped during the course of the present study, as discussed below.

The study provides an important milestone in the development and application of methodologies to support WRTM. It brings WRTM applications into the mainstream of network modeling and simulation tools, and demonstrates the potential of both WRTM for urban areas and states, as well as of TrEPS tools to evaluate and develop strategies on an ongoing basis, as part of the routine functions of planning and operating agencies.

In addition to achieving the intended study objectives, several contributions and new developments were accomplished as part of the study. These include: (1) building a library of calibrated traffic flow relations under inclement weather for different areas of the United States (East and West coasts, Midwest, Mid-Atlantic, Mountain region); (2) development of a prototype scenario manager, intended to facilitate application of TrEPS for different types of weather and other scenarios; (3) development of *key performance indicators* (KPI's) to evaluate the effectiveness of particular WRTM strategies in a given network, allowing the model user/agency personnel to compare network performance overall as well as for particular portions of the network, O-D pairs or user segments, with and without WRTM as well as for different WRTM strategies; (4) the concept of "equivalent demand reduction" needed to offset network performance impairment introduced by particular inclement weather conditions, and maintain level of service expected under normal weather conditions; and (5) a novel deployment model whereby TrEPS can be initially introduced and maintained as a remote service hosted and maintained by a different organization for as-needed access.

Many important findings were reached through this study regarding the role that network models and simulation methodologies can play in the further development and deployment of WRTM strategies, and the process through which such tools could be most effective in helping agencies attain their objectives within available resources. Several recommendations are provided regarding additional steps to build on the findings and accomplishments of the study to further evolve the TrEPS methods into an integrated platform for WRTM, and advance the state of the art and practice of WRTM. These are categorized into immediate steps, which build up in a direct manner on the completed work, and medium term steps geared towards a more complete methodological capability for WRTM in the context of system management activities. Immediate steps include: (1) Implementation-driven development of the scenario manager prototype developed in the present study to support TrEPS deployment and application, primarily in conjunction with WRTM preparedness in response to near-term forecasts of impending inclement weather; (2) Test deployment along the remote-hosted model established in the present study, covering a longer duration that would allow a certain number of actual inclement weather instances; and (3) In conjunction with the above deployment, it would be important to conduct a behavior tracking study that would allow observation of actual user responses to WRTM strategies, with particular focus on demand management strategies.

Steps over the medium term to improve the methodological basis of existing TrEPS methodology, particularly with regard to expanding the range of its usefulness to a more comprehensive scope of WRTM activities, include: (1) Integrate accident response functionality with WRTM in the real-time TrEPS platform; (2) integrate fleet routing functionality, e.g. for snow removal equipment, preventive sanding and freeze-melting agent spreading, and other logistical processes, with the TrEPS platform; (3) incorporate transit-related capabilities to provide essential functionality in larger metropolitan areas with substantial reliance on transit services, or in smaller-sized areas that wish to take advantage of the additional mobility provided by transit during weather-related disruptions; (4) Revisit O-D estimation aspects in the on-line TrEPS context, and test these in connection with an active scenario manager that can retrieve calibrated a priori demand matrices for the particular weather scenarios under consideration; (5) Enhance the ability of the TrEPS simulation tools to assess the impact on relative safety of inclement weather, and correspondingly the impact of WRTM measures on that important system performance indicator; and (6) Incorporate mobile data in future deployment-based development and testing of the methodology.

1. Introduction

1.1 Background

Weather events such as precipitation, fog, high winds and extreme temperatures cause low visibility, slick pavement, reduced roadway capacity and other hazardous conditions on roadways. The disruptive effect of inclement weather on traffic has staggering impact on safety- about 28% of all highway crashes and 19% of all fatalities involve weather-related adverse conditions as a factor. Additionally, adverse weather accounts for about 25% of delays on freeways due to reduced service capacity (often at the most critical of times), diminished reliability of travel, and greater risk of accident involvement. To mitigate the impacts of adverse weather on highway travel, the Federal Highway Administration (FHWA) Road Weather Management Program (RWMP) has been involved in research, development and deployment of weather responsive traffic management (WRTM) strategies and tools. Dealing with adverse weather requires not only sensing of traffic conditions, but also the ability to forecast the weather in real-time for operational purposes. Recognizing the importance of tying weather and traffic management together in areas exposed to extreme weather situations, such as hurricanes and floods, some TMC's such as the Houston TranStar TMC co-locate the weather service personnel with the usual traffic management agencies (police, traffic operators, EMS). The most ambitious initiative in this regard is the *Clarus* weather system, intended to provide traffic management centers with accurate real-time weather information *(Pisano and Goodwin, 2002; Mixon-Hill Inc. et al., 2005; Pisano et al., 2005; and FHWA Clarus web site)*. Weather information, along with roadway traffic information obtained from ITS sensors, enable promising opportunities to improve traffic operations and management under inclement weather.

In order to reduce the impacts of inclement weather events and prevent congestion before it occurs, weather-related advisory and control measures could be determined for predicted traffic conditions consistent with the forecast weather, that is, anticipatory road weather information. This calls for integrated real-time WRTM and a Traffic Estimation and Prediction System (TrEPS). Because the dynamics of traffic systems are complex, many situations necessitate strategies that anticipate unfolding conditions instead of adopting a purely reactive approach. Real-time simulation of the traffic network forms the basis of a state prediction capability that fuses historical data with sensor information, and uses a description of how traffic behaves in networks to predict future conditions, and accordingly develop control measures *(Jayakrishnan et al. 1994; Mahmassani 1998, 2001)*. The estimated state of the network and predicted future states are given in terms of flows, travel times, and other time-varying performance characteristics on the various components of the network. These are used in the on-line generation and real-time evaluation of a wide range of measures, including information supply to users, VMS displays, coordinated signal timing for diversion paths, as well as weather-related interventions (through variable speed limits, advisory information, signal timing adjustments and so on).

In a previous FHWA project, a methodology for incorporating weather impacts in Traffic Estimation and Prediction Systems (TrEPS) is developed *(Mahmassani et al., 2009)*. The project addressed both supply and demand aspects of the traffic response to adverse weather, including user responses to various weather interventions such as advisory information and control actions. The methodology was incorporated and tested in connection with the DYNASMART-P simulation-based DTA system, thereby providing a tool for modeling the effect of adverse weather on traffic system properties and performance, and for supporting the analysis and design of traffic management strategies targeted at such conditions.

The purpose of the current project is to further calibrate and validate the methodological development made in the previous project to advance the state of practice of weather-responsive traffic management. The weather-sensitive on-line TrEPS can serve as a catalyst for the development and advancement of effective WRTM strategies, as it allows TMC's to test and evaluate various site-specific traffic control/advisory plans. Essential steps to this end include having DYNASMART-P (off-line) fully calibrated to local traffic and weather conditions and seamlessly extending its functionalities to DYNASMART-X (on-line) for the real-time operations.

1.2 Integration of Weather in TrEPS

A real-time traffic estimation and prediction system (TrEPS) is an essential methodology to enable implementation and evaluation of on-line traffic management, as it can incorporate field observations and traffic measures, as well as estimate and predict network states. DYNASMART-X *(Mahmassani et al., 1998; Mahmassani and Zhou, 2005)* and DynaMIT-R *(Ben-Akiva et al., 2002)*, both developed largely under FHWA support, use a simulation-based dynamic traffic assignment (DTA) approach for real-time traffic estimation and prediction. As a deployable real-time system, TrEPS must recognize the fact that OD demand information and network conditions can only be reliably available for a short period of time in the future. One way to account for the uncertainty of future information is the rolling horizon (RH) approach *(Peeta and Mahmassani, 1995)*. In a RH framework, new OD desires are being continuously estimated and corrected using the inflow of actual observations from different data sources. Based on the updated OD demand, every prediction stage predicts a new network state. With every roll, the newly estimated variables overwrite the ones obtained from the previous stage, i.e. only the most up-to-date information is used.

As a state-of-the-art real-time TrEPS, DYNASMART-X interacts continuously with multiple sources of real-time information, such as loop detectors, roadside sensors, and vehicle probes, which it integrates with its own model-based representation of the network traffic state. The system combines advanced network algorithms and models of trip-maker behavior in response to information in an assignment-simulation-based framework to provide, in real-time: (1) estimates of network traffic conditions, (2) predictions of network flow patterns and travel times over the near and medium terms in response to various contemplated traffic control measures and information dissemination strategies, and (3) anticipatory traveler and routing information to guide trip-makers in their travel *(Dong, Mahmassani, and Lu, 2006)*. The system includes several functional modules (e.g. OD estimation, OD prediction, real-time network state simulation, consistency checking, updating and resetting functions, and network state prediction), integrated through a flexible distributed design that uses CORBA (Common Object Request Broker Architecture) standards, for real-time operation in a rolling horizon framework with multiple asynchronous horizons for the various modules *(Mahmassani et al., 2004)*. The functionality of DYNASMART-X is achieved through judicious selection of modeling features that achieve a balance between representational detail, computational efficiency and input data requirements. Further detail on the structure and components of a TrEPS such as DYNASMART-X is available in the appropriate manuals.

In the previous FHWA project *(Mahmassani et al., 2009)*, the principal supply-side and demand-side elements affected by adverse weather were systematically identified and modeled in the TrEPS framework. The models and relations developed were calibrated using available observations of traffic and user behavior in conjunction with prevailing weather events. The proposed weather-related features have been implemented in DYNASMART, and demonstrated through successful application to a real world network, focusing on two aspects: (1) assessing the impacts of adverse weather on transportation networks; and (2) evaluating effectiveness of weather-related advisory/control strategies in alleviating traffic congestion due to adverse weather conditions. The procedures implemented provide immediately applicable tools that capture knowledge accumulated to date regarding weather effects on traffic. The application to a real world network shows that the proposed model can be used to evaluate weather impacts on transportation networks and the effectiveness of weather-related variable message signs and other strategies.

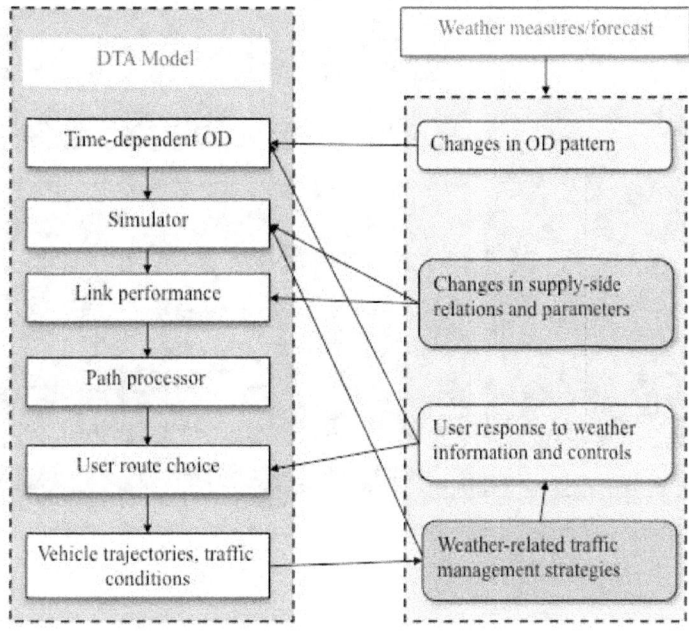

Figure 1-1. Incorporating Weather in DTA Model (Source: Mahmassani et al., 2009)

In addition, the modular structure of the system enables consideration of multiple future scenarios simultaneously, as illustrated in the GUI snapshot in the Figure below, for the Maryland CHART network (along I-95 between Washington, D.C. and Baltimore). In the left pane, the estimated traffic conditions are shown, in a manner that is completely synchronized with real time; i.e. it displays currently prevailing conditions, as seen by the model. In the right panes are displayed prediction results, using P-DYNA. Let's say that adverse weather has been anticipated, and this has been communicated to the TrEPS. A prediction is then generated for the traffic under that traffic scenario, which would be viewed as the base case (using P-DYNA0). To evaluate the effectiveness of an intervention, say the display of various advisory messages, and dissemination of information through the internet and mass media, another scenario can be run in parallel, using another copy of P-DYNA (say P-DYNA1), to predict conditions with the intervention. Comparing the results of P-DYNA0 vs. P-DYNA1 would then allow the traffic manager to decide accordingly. This feature of DYNASMART-X, developed for the Maryland CHART network *(Mahmassani et al., 2005)*, enables parallel execution of several alternative intervention scenarios in the context of real-time decision support for traffic management. Of course, various comparative statistics can also be displayed through the GUI.

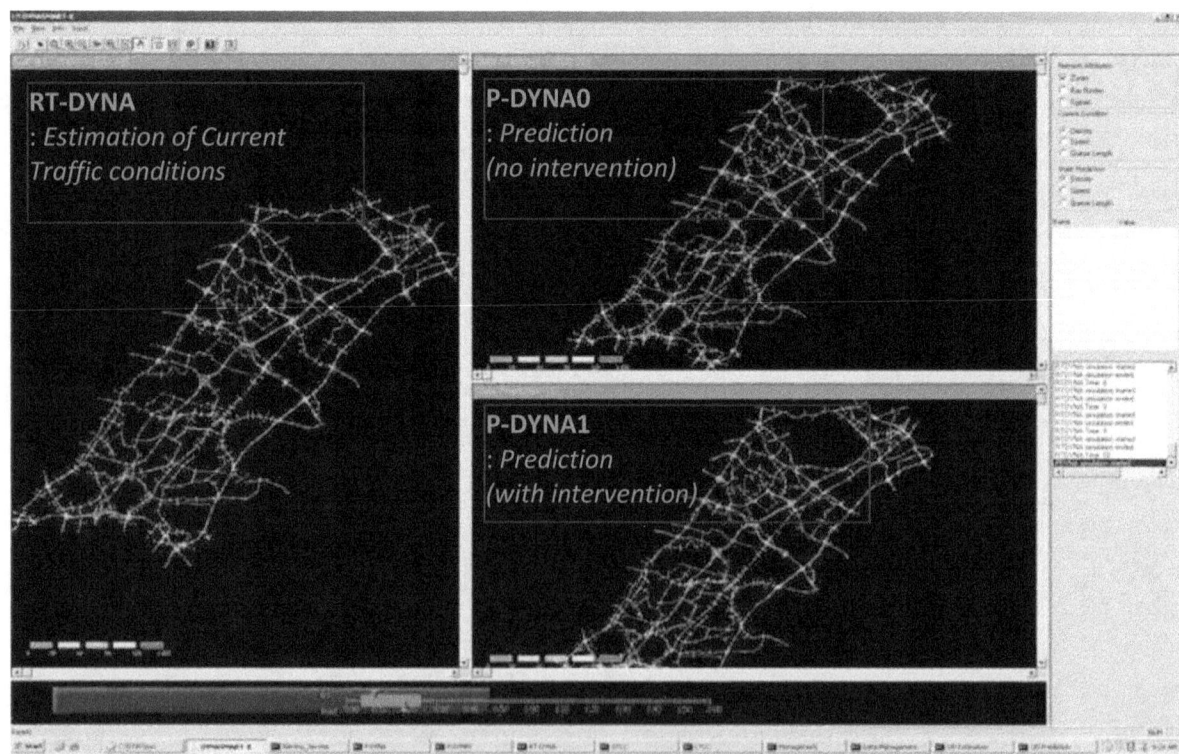

Figure 1-2. DYNASMART-X, Real-time Traffic Estimation and Prediction System

1.3 Weather Responsive Traffic Management Strategies

1.3.1 Advisory Information

Road weather information, such as en route weather warning and route guidance, can be disseminated through radio, internet, mobile devices, roadside VMS and so on. Weather warning VMS have been implemented in the field, and shown to be effective in decreasing the average speed as well as the variance in speed, and hence helpful in increasing safety and reliability for the traveling public *(Luoma et al., 2000; Rämä, 2001)*. Weather VMS also proved most effective when adverse weather and road conditions were not easy to detect. Weather advisory VMSs, in the form of slippery road condition sign and fog (low visibility) sign, have been implemented and tested in Europe. For example, in Finland slippery road condition sign, implemented in combination with the minimum headway sign, decreased the mean speed by 1.2 km/h with the steady display and by 2.1 km/h when the sign was flashing *(Rämä, 2001)*. Hogema and van der Horst *(1997)* showed that the Dutch fog warning signs, implemented in conjunction with variable speed limits, decreased the mean speed in fog by 8 to 10 km/h (i.e. 5 to 6 mph). Cooper and Sawyer *(1993)*, on the other hand, found that the automatic fog-warning system on the A16 motorway in England reduced the mean vehicle speed by approximately 3 km/h (i.e. 2 mph). A comprehensive synthesis of recent developments and applications focusing on US practice is presented in a FHWA report (Gopalakrishna et al, 2011).

1.3.2 Control Strategies

In addition to weather warning and advisory strategies, control methods could also be deployed for enforced traffic management during inclement weather events. Variable speed limits (VSL) and weather-responsive signal controls are two examples of such applications.
VSL utilizes traffic speed and volume detection, weather information, and road surface condition technology to determine appropriate speeds at which drivers should be traveling, given current roadway and traffic conditions. These advisory or regulatory speeds are usually displayed on overhead or roadside variable message signs (VMS). VSL systems are already being used as part of incident management, congestion management, weather advisory, or motorist warning system to help potentially to enhance the safety and reliability of roadways *(Robinson, 2000)*. VSL are sometimes displayed alongside the weather advisory VMS to inform travelers as well as enforce traffic safety, such as on the interurban Highway E18 in Finland *(Rämä, 1999)*. The speed limit posted could be adjusted based on prevailing weather conditions according to a look-up table. For instance, on E18 in Southern Finland between Kotka and Hamina speed limits are set as 120 km/h (74 mph) for good road conditions; 100 km/h (62 mph) for moderate road conditions; and 80 km/h (49 mph) for poor road conditions *(Rämä, 1999)*. Similarly, on I-90 Snoqualmie Pass in Washington State, the posted speed limit is reduced from 65 mph, in ten-mph increments, to 35 mph, depending on visibility and severe weather, itself obtained from multiple weather stations, snow plow operators, and State Patrol *(Robinson, 2000)*. Variable speed limits could also be determined based on visibility, friction and the prevailing traffic conditions. For example, on Interstate 80 in Nevada, speed limits (in ten-mph increments) are computed

using a logic tree based on the 85th percentile speed (measured using speed loops), visibility (collected from visibility detector), and pavement conditions (frost, ice, rain, or dry conditions from the Road Weather Information System weather station) *(Robinson, 2000)*. On the New Jersey Turnpike, the posted speed limits can be reduced from the normal speed limit (depending on the milepost location 65 mph, 55 mph, and 50 mph) in five-mph increments, to 30 mph, based on average travel speed collected from inductive loop detectors *(Robinson, 2000)*. On the E6 Halland motorway in Sweden, the speed limits are controlled according to the expected friction coefficient on the road. Specifically, if the friction is expected to be 0.4 (moderate rain, light snowfall) the speed limit will be set to 110 km/h; friction 0.3 (heavy rain, moderate snowfall) results in 100 km/h; friction 0.2 (very heavy rain, heavy ice formation) results in 80 km/h; and the extremely low friction 0.1 (cloudburst, very heavy ice formation) results in 60 km/h. The expected friction coefficient is calculated based on temperature, moisture, wind speed and wind direction *(Lind, 2007)*.

Since weather events can reduce the effectiveness of traffic signal timing plans designed for use in clear, dry pavement conditions, a weather-responsive signal timing plan is desired when adverse weather events occur. A few studies reported empirical observations on the effect of adverse and extreme weather conditions on signal timing input parameters. Agbolosu-Amison et al. *(2004)* investigated the effect of inclement weather on start-up lost time and saturation headway at a study site in northern New England. The study reveals that inclement weather has a significant impact on saturation headways (the highest increases in average saturation headway of 21%), particularly once slushy conditions start. Maki *(1999)* reported a 40% reduction in average speed, 11% reduction in saturation flow rate, and 50% increase in start-up lost time due to adverse weather conditions at an arterial corridor in the Minneapolis–St. Paul Twin Cities metropolitan area of Minnesota. Perrin et al. *(2001)* investigated the change in traffic flow parameters under various weather severity levels at two intersections in Salt Lake City throughout the winter of 1999–2000. According to this study, the largest decrease in vehicle performance occurs when snow and slush begin to accumulate on the road surface. Saturation flows decrease by 20%, speeds decrease by 30%, and start-up lost times increase by 23%.

Along with studies that observe the effect of weather on signal timing parameters, empirical or simulation studies have been conducted to determine optimal signal plans during inclement weather. Lieu and Lin *(2004)* assessed the benefits of retiming signal control under adverse weather conditions using traffic simulation. The study considers a simple numerical example that involves a hypothetical arterial corridor with four successive intersections and a single set of signal timing input parameters for an adverse weather scenario. The study found that potential operational benefits of retiming signals can be realized only when traffic flows are moderately high. Maki *(1999)* performed field tests in an arterial corridor in the Minneapolis–St. Paul Twin Cities metropolitan area to evaluate the feasibility of implementing a coordinated traffic signal timing plan that will accommodate traffic under adverse weather. Using field data on weather impacts and the Synchro signal optimization software, the study concluded that the "corridor operation was not radically affected by the adverse weather"; this is mainly "due to the fact that there are fewer vehicles to cause delay to during bad weather" Maki *(1999)*. Agbolosu-amison et el. *(2005)* designed and conducted several simulation experiments to understand the impact of different factors affecting the magnitude of the operational benefits of special timing plans for inclement weather. Two signalized arterial corridors were selected as case studies, and optimal signal plans were developed for six different weather and road surface conditions for each corridor by using four different simulation models *(TRANSYT-7F, Synchro, CORSIM, and SimTraffic)*. To develop the weather-specific models, the saturation flow rate and free flow speed corresponding to each weather and road surface condition were coded by using the reduction factors, which gave the percent reduction relative to the dry condition rate. The results show that signal retiming during inclement weather can result in significant operational benefits (as high as a 20% reduction in control delay in some cases). Al-Kaisy and Freedman *(2006)* present a set of recommended guidelines that relate weather conditions to operational impacts and potential benefits of weather-responsive signal timing through a systematic investigation considering isolated and coordinated signalized intersections in urban and suburban areas under various traffic conditions. Both operational and safety analyses were conducted in this investigation. Traffic signal optimization and microscopic traffic simulation were used to perform the operational analysis with average travel time as a performance measure. The adequacy of change and clearance intervals and the presence of dilemma zones were used as safety indicators at signalized intersections in performing the safety analysis.

In practice, Goodwin *(2003)* and Goodwin et al. *(2004)* presented two case studies of weather-responsive signal control operations, which are intended to issue traffic signal preemption (e.g., to clear traffic from a beach or drawbridges) or to slow the overall intersection progression speed in response to poor road weather conditions. By selecting signal timing plans based upon prevailing weather conditions traffic managers can improve roadway mobility and safety. A description of weather-related parameters in simulation models and the benefits of weather-responsive signal timing are also discussed.

The above discussion of prior work reveals that efforts to devise weather-related traffic management systems have remained limited to a few countries and locales, though recognition for the need for such intervention continues to increase. Furthermore, the need for and potential usefulness of a weather-enabled TrEPS presents a significant though challenging opportunity for advancing the state of practice.

1.4 Conceptual Framework for On-line Implementation

1.4.1 Implementation of WRTM strategies using TrEPS models

To effectively manage the flow of traffic during inclement weather conditions, many agencies implement a wide variety of WRTM strategies. In general, based on pre-defined operational procedures for different weather types and severities, corresponding strategies are employed in response to prevailing weather conditions. Because the dynamics of traffic systems are so complex, however, WRTM strategies selected based on such general rules may not always perform as expected. This calls for integration of WRTM and a real-time Traffic Estimation and Prediction System (TrEPS), which allows incorporating predicted traffic conditions under different strategies into the decision of appropriate WRTM strategies. The real-time TrEPS model interacts continuously with loop detectors, roadside sensors and vehicle probes, providing real-time estimates of traffic conditions, network flow patterns and routing information. Based on the current network state, a prediction is then generated for the traffic under future weather conditions and weather-related interventions providing the predicted effect of WRTM strategies on the real world network. An overall framework of the implementation of WRTM strategies using the weather-responsive TrEPS model is presented in Figure 1-3. The framework comprises three components: WRTM Strategy Repository, Scenario Manager and DYNASMART-X. The WRTM Strategy Repository contains a set of available WRTM strategies defined for different weather situations. Based on existing guidelines and practices adopted by local operating agencies, several alternatives could be identified and included in each weather category. For example, when the rain intensity exceeds a certain threshold, different combinations of individual advisory/control method (e.g., VMS, VMS + speed limit, and VMS + signal timing) might be considered as available implementation options. In case of a snowfall, decisions might involve choosing between different routing and scheduling options for snow plow operations. When the Scenario Manager receives the prevailing weather conditions and the future weather information, it firsts generates the weather scenario input file (i.e., weather.dat) for the next prediction horizon that will be simulated in DYNASMART-X. Next, it retrieves available WRTM strategies based on the specified weather condition from the WRTM Strategy Repository. Users might choose two or more strategies under consideration. The Scenario Manager then creates a set of input files for each strategy (e.g., VMS.dat, WorkZone.dat, control.dat, etc.) and supplies them to DYNASMART-X along with the weather scenario file. In DYNASMART-X, based on the estimated current traffic conditions using real-time traffic surveillance data, the future traffic conditions are predicted for different scenarios. The predicted network performance measures produced under different intervention scenarios will allow the traffic manager to evaluate effectiveness of each strategy and select the best WRTM strategy for the current situation.

The weather-responsive TrEPS model would also help decision-making in modifying plans for various roadside events such as road construction, pavement works and planned special street events. When such events encounter unexpected adverse weather conditions, the traffic manager can simulate different weather and traffic management scenarios to assess the impact of weather conditions on traffic and decide how to modify the current plan to minimize the congestion and risk of accident.

1.4.2 Evaluation of WRTM strategies using TrEPS models

Evaluating effectiveness of various WRTM strategies would require several times of implementation and measurement. As we observe only the outcome associated with selected WRTM strategy, it would take time until we have sufficient number of similar occasions for which different scenarios are tested and outcomes are collected. In this case, historical data and past experiences need to be used to assess the performance of the selected strategy. The evaluation procedure can also be facilitated by the use of the TrEPS model framework. Figure 1-4 presents the post-process of the real-time WRTM implementation in the context of the same framework shown in Figure 1-3. After applying the selected WRTM strategy, DYNASMART-X obtains the traffic surveillance data and estimates the resulting network state. This can be viewed as the network performance outcomes produced under the implemented WRTM strategy and used for the traffic manager to judge its effectiveness. If it is considered necessary to modify/discard the selected strategy or add a new strategy, the Scenario Manager will help update the WRTM Scenario Repository accordingly. The updated strategies for the experienced weather situation are stored in the repository and will be retrieved on demand next time the similar weather event occurs.

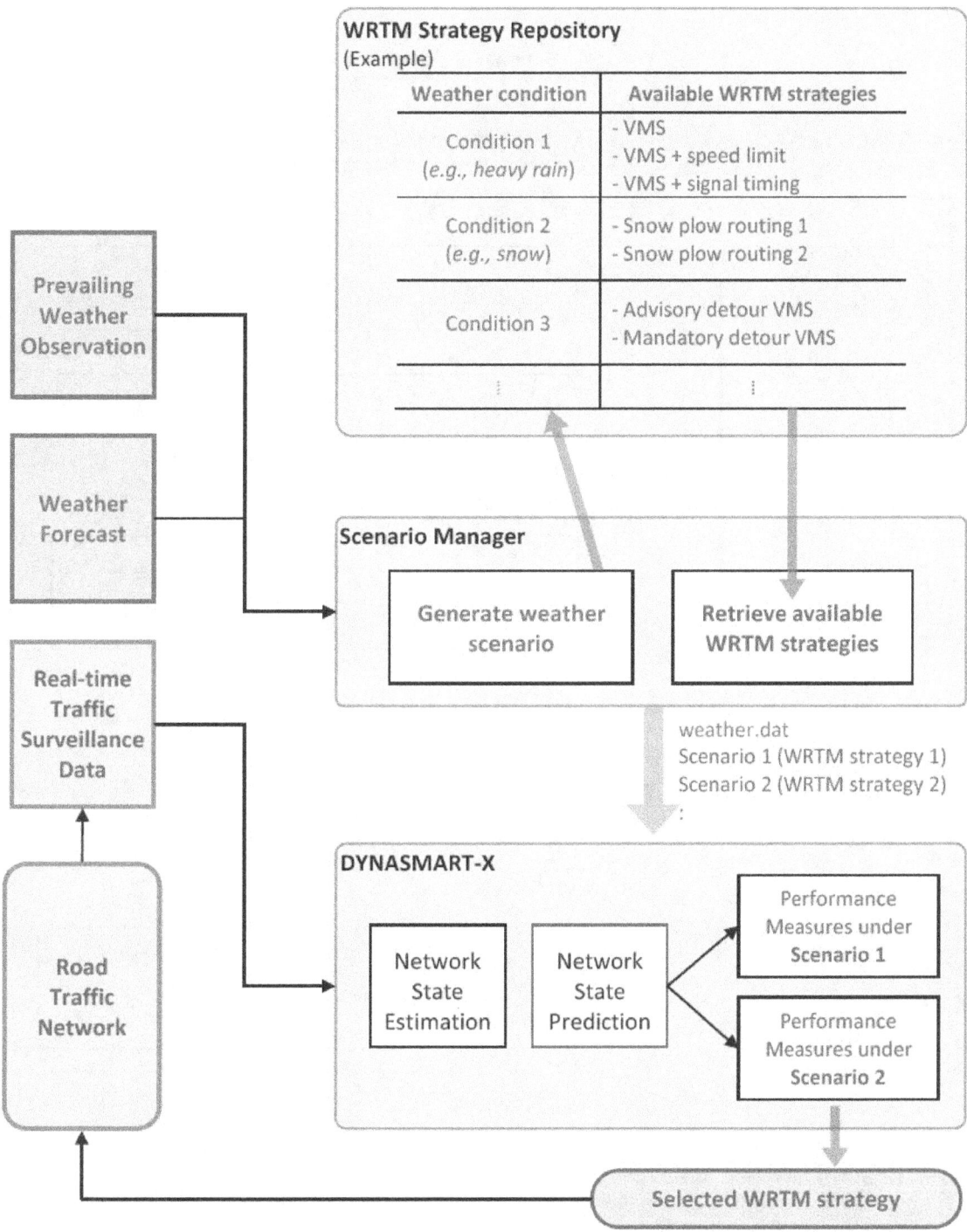

Figure 1-3. Framework for Implementing WRTM strategies using TrEPS models

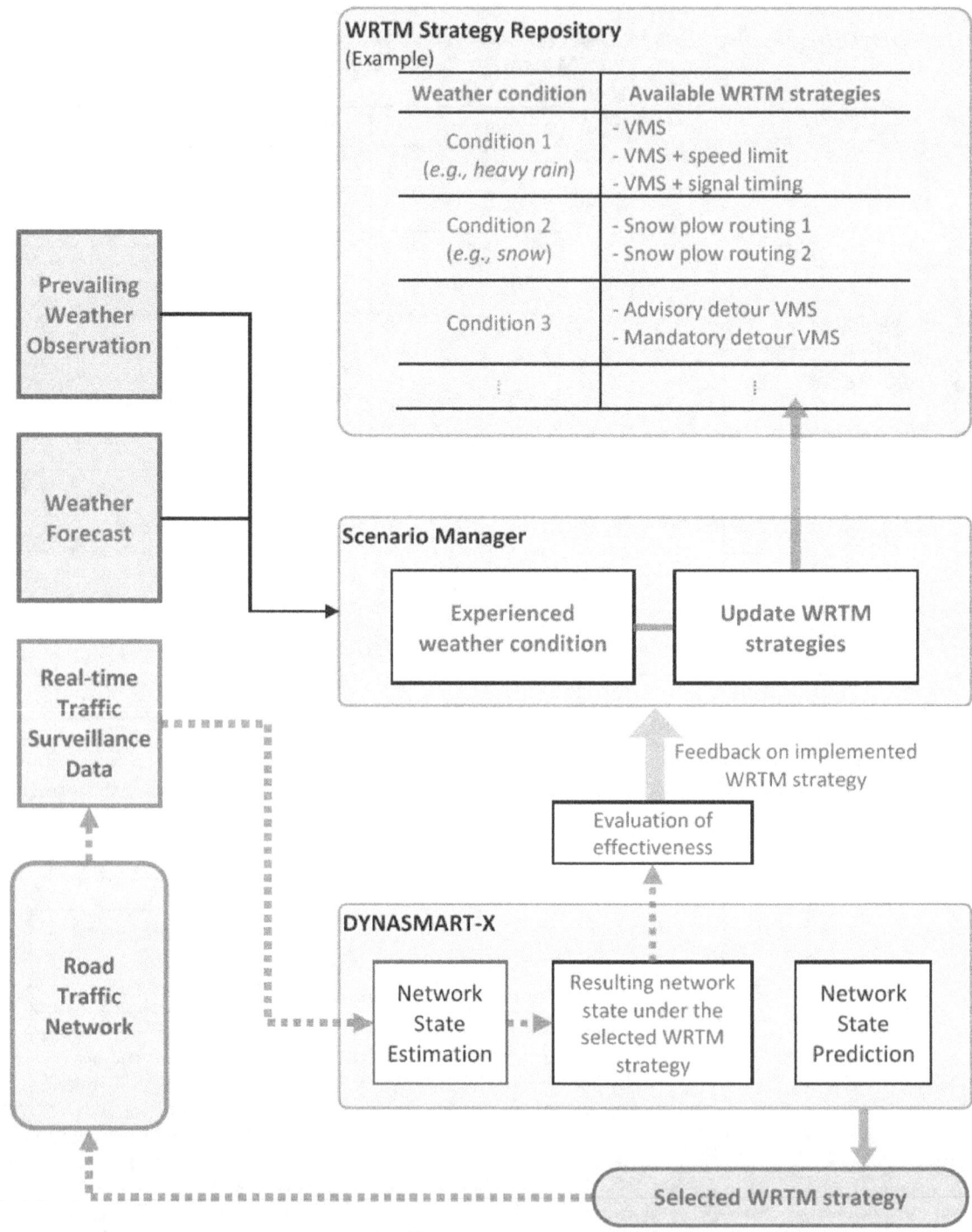

Figure 1-4. Framework for Evaluating WRTM strategies using TrEPS models

1.5 Project Approach

Adverse weather conditions have a significant impact on traffic conditions by causing direct and indirect changes in the roadway environment. Direct changes include reduced pavement friction due to rain, snow and ice; and low visibility due to fog and heavy rain/snow. Indirect changes are the cases where roadway service capacity is reduced because users tend to drive at lower speed during inclement weather; or traffic volume is changed because of travelers' departure time shift and trip cancelation. Both direct and indirect impacts greatly increase risk of accident, delays and uncertainty in travel time. To mitigate the impacts of adverse weather on highway travel, the Federal Highway Administration (FHWA) Road Weather Management Program (RWMP) has been involved in research, development and deployment of *weather responsive traffic management (WRTM)* strategies and tools. In recent years, much effort has been made under this program to encourage Transportation Management Center (TMC) to integrate weather information in traffic operations and to develop various WRTM strategies and evaluation guidance to deal with adverse weather *(Cluett et al., 2011; Gopalakrishna et al., 2011).*

The overall goal of the study is to implement and evaluate weather responsive traffic management strategies using TrEPS models. The project shows the value of incorporating weather effects in traffic modeling software, and how combining weather forecasts with traffic prediction can help local and state agencies in weather-related traffic management. To accomplish this goal, the team performed the following tasks:

- Review and summarize existing knowledge on the application of weather-responsive advisory and control strategies such as dynamic message signs, 511, variable speed limits, and signal control.
- Identify and summarize candidate networks that have already been calibrated for normal conditions and can be further developed to account for weather impacts.
- Calibrate and validate the TrEPS models using observed weather and traffic data for various locations in the US and under different weather conditions.
- Implement and evaluate weather responsive traffic advisory and control strategies at the corridor and network levels utilizing the TrEPS models that account for traffic response to inclement weather.

Figure 1-5 depicts the key steps followed in this project for implementation and evaluation of weather-responsive TrEPS system.

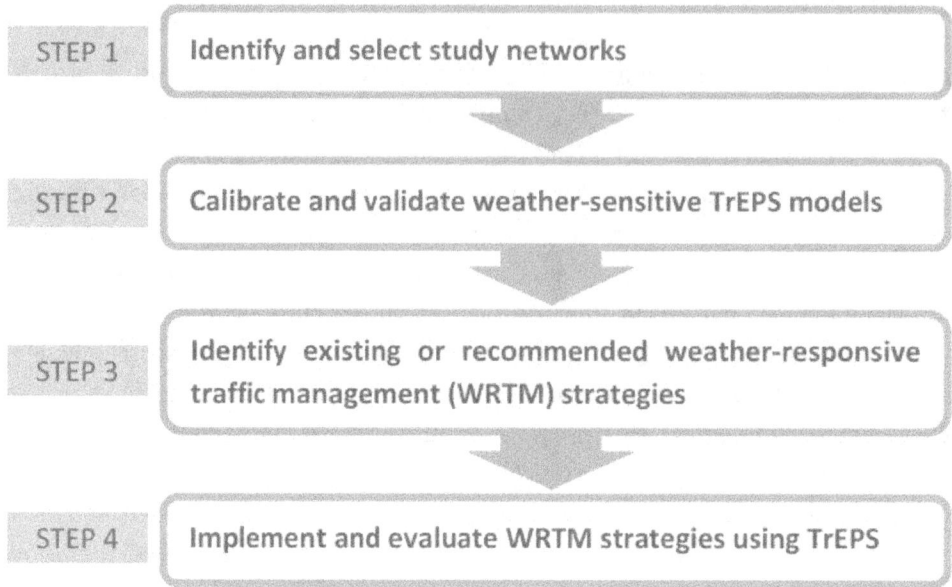

Figure 1-5. Key Steps for Implementation and Evaluation of Weather Responsive TrEPS System

1.6 Structure of Final Report

This report discusses the activities conducted as part of Implementation and Evaluation of Weather Responsive Traffic Estimation and Prediction System study. The following chapters are covered:
- Chapter 2 describes the procedure for selecting four study sites among major U.S. cities to conduct the calibration and validation of TrEPS models. For the selected cities, the chapter presents how the networks for the simulation-based DTA model are built and the sources of traffic and weather data for each network.
- Chapter 3 describes the calibration and validation of weather-sensitive TrEPS model. Detailed procedures and results for calibrating the supply-side parameters, i.e., traffic flow model parameters and weather adjustment factors, and the demand-side parameters, i.e., time dependent OD matrices for the simulation analysis, are presented.
- Chapter 4 reviews a list of WRTM strategies that are found in the literature and discusses the project team's activities for identifying existing WRTM strategies used by the agencies in the selected four study sites. Based on the identification, the chapter provides a summary of recommended WRTM strategies for each network for the implementation.
- Chapter 5 provides an overview of evaluation approaches to assess the benefits of WRTM strategies and various performance measures that can be used in the evaluation process. Detailed discussion and examples on mobility performance measures are presented.
- Chapter 6 describes the implementation and evaluation of selected WRTM strategies for study networks. For the three networks that perform the weather-responsive traffic management, which include Chicago, Long Island and Salt Lake City, procedures for coordinating with local agencies and detailed analysis results are presented.

- Chapter 7 presents the conclusions, including lessons learned and recommendations for next steps needed to advance the state of the art and of the practice of TrEPS use for WRTM.

2. Study Networks

2.1 Site Selection Procedure

This chapter presents the procedures and results for the selection of the study sites. We developed a set of criteria to measure appropriateness of each candidate network in the application of the weather-sensitive TrEPS models. Based on a systematic evaluation process, four study sites are recommended by the project team and approved by FHWA for the calibration and validation of the TrEPS model. The four sites include: Chicago, IL; Salt Lake City, UT; New York, NY; and Irvine, CA.

2.1.1 List of Candidate Sites

Initially, the project team was considering the following eight networks as candidate sites for calibrating the weather-sensitive TrEPS model.
- Irvine, CA
- Portland, OR
- Baltimore, MD
- New York, NY
- Houston, TX
- Chicago, IL
- Virginia Beach, VA
- Salt Lake City, UT

Figure 2-1 presents the spatial distribution of the above-mentioned sites and corresponding geographical divisions based on U.S. standard regions for temperature and precipitation. As shown in the figure, the eight networks provide good geographic representation across the U.S.

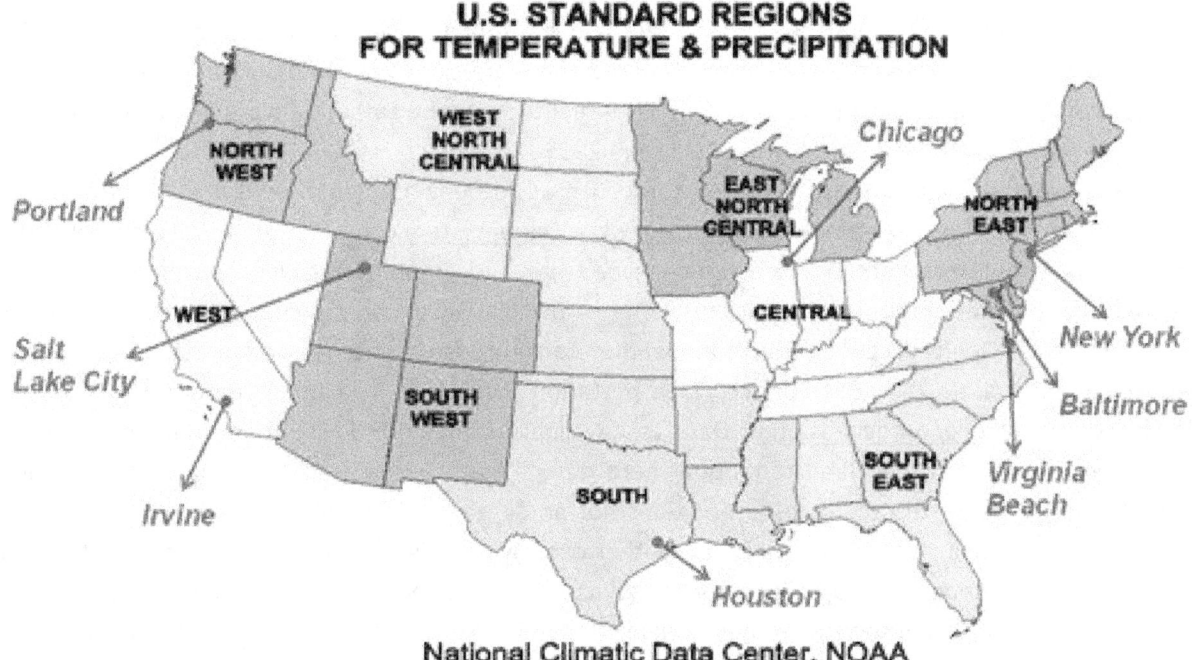

Figure 2-1. Geographical Distribution of Candidate Sites (Source: National Climatic Data Center, NOAA)

2.1.2 Criteria for Network Selection

2.1.2.1 Traffic and Weather data

Table 2-1 primarily presents the availability of traffic and weather data for each network; all networks have available sources of such data and a detailed discussion is provided as follows.

Traffic Data

In general, networks that are connected with a web-based data archive system such as PeMS, PORTAL and CATT Lab provide a wide range of options for time periods and data resolution, as well as easy access to the data. In terms of temporal resolution of traffic data, 5 minutes is considered to be the most appropriate interval for off-line calibration of the simulation-assignment component since currently the most readily available source of weather data have a resolution of 5 minutes. However, for eventual actual on-line operation of the TrEPS, it would be desirable to have finer resolution of the traffic flow data stream, preferably not to exceed one minute.

Weather Data

In terms of weather data, Automated Surface Observing System (ASOS) stations and Environmental Sensor Station (ESS) in the *Clarus* system are considered as major sources and examined in detail.

ASOS : ASOS stations located in adjacent airports of each study site are presented in **Appendix A**. Different numbers of stations are available for each network, but the coverage depends on the size of the network. For example, as shown in **Appendix A**, although there is only one ASOS station for both

Irvine and Baltimore, the Irvine network completely falls within 10 miles from the station whereas the Baltimore network is only partially covered by the same range. Based on current information, Irvine, Portland and Chicago show the best coverage of weather data from adjacent ASOS stations unless other sources of weather data are further provided. Five-minute surface weather data are available since 2000 for all networks except Irvine, which has data from 2005.

<u>*Clarus*</u> : Weather data are also collected through roadside environmental sensor stations (ESS) and available on the *Clarus* system website (http://www.clarus-system.com). ESS data related to weather conditions include precipitation (occurrence, type, rate and amount), visibility, pavement condition (snow depth, water depth and ice thickness) and so on. Clarus system provides two different ways of accessing data. One is access to historical weather data through an active archive established by The University of North Dakota Surface Transportation Weather Research Center (UNC STWRC, http://stwrc.und.edu/ undclarus.php). Data are available from 2008-12-31 to present at 20-minute intervals. The interval (i.e., time resolution) here differs from that of traffic data in that it indicates simply the frequency of recording observations not an aggregate interval for which observations are averaged. The other is access to real time weather data by applying for subscription. Starting from the time of the subscription, data collected from selected ESS are recorded and stored in the system at every pre-defined time interval, which is available for 5, 10, 15, 20, 30 and 60 minutes. The latter case would be useful for on-line operation of the TrEPS.

For off-line calibration using historical weather data, ASOS data seem to be more suitable than Clarus because more observations are available for longer time periods and more timestamps. For networks with poor ASOS coverage, however, the Clarus database can be used as a good complement to ASOS data since typically ESSs are installed in more sites along roadsides while ASOSs are only found in airports. Locations of both types of weather stations are compared in Appendix C. Irvine and Houston networks are not presented because Clarus data are not available for those sites. Currently no ESS is found near Irvine network although many other places in California are available in the system whereas Texas itself is not included in the contributor list of Clarus system. Networks that might benefit most from using Clarus data include Baltimore, New York, Salt Lake City and Portland since ASOS and ESS together provide far better coverage than ASOS does alone for these sites as shown in **Appendix B**. On the contrary, there are very few ESSs available for Virginia Beach and Chicago.

Table 2-1. Characteristics of Candidate Networks (Traffic and Weather Data)

Criteria			Irvine, CA	Portland, OR	Baltimore MD	New York, NY	Houston, TX	Chicago, IL	Virginia Beach, VA	Salt Lake City, UT
Traffic Data		Historical detector data availability	Yes	Yes	Yes	Partial	Partial	Yes	Yes	Yes
		Source of data	PeMS (pems.eecs.berkeley.edu)	PORTAL (portal.its.pdx.edu)	CATT Lab (www.cattlab.umd.edu)	NYDOT, CTDOT, NJDOT websites	TranStar, TTI	GCM	ADMS Virginia (adms.vdot.virginia.gov)	Wasatch Front Regional Council
		Time period	≥ 5 yrs	≥ 5 yrs	≥ 5 yrs	Through 2008 (NY), 2007 (CT), 2006 (NJ)	One month in 2006	2004 – 2008	≥ 5 yrs	≥ 3 yrs
		Data resolution	5min, 1hr	5min, 15min, 1hr	5min, 15min	5, 10 and 15 min	30 sec.	5min	5min, 15min	5min
Weather Data		Historical weather data availability	Yes	Yes	Yes	Yes	Yes	Yes	Yes	Yes
	ASOS	# of stations	1	4	1	9	3	5	1	1
		Time period	2005 - present	2000 - present	2000 - present	2000 - present	2000 - present	2000 - present	2000 - present	2000 - present
		Data resolution	5min	5min	5min	5min	5min	5min	5min	5min
	Clarus (ESS)	# of Stations	NA*	5-10	10-15	10-15	NA**	1-5	1	1-5
		Time period	-	2008-12-31 - present	2008-12-31 - present	2008-12-31 - present	-	2008-12-31 - present	2008-12-31 - present	2008-12-31 - present
		Data resolution	-	20min	20min	20min	-	20min	20min	20min
		Other sources							ADMS Virginia	

* No Environmental Sensor Station (ESS) is found in this area at the time of the analysis
** Clarus system is not available for Texas at the time of the analysis.

2.1.2.2 Geographic Scope and Weather pattern

In Table 2-2, eight cities are compared based on a geographic scope, an extreme weather pattern and an average weather pattern, respectively.

Table 2-2. Characteristics of Candidate Networks (Geographic Scope and Weather Pattern)

Criteria		Irvine, CA	Portland, OR	Baltimore, MD	New York, NY	Houston, TX	Chicago, IL	Virginia Beach, VA	Salt Lake City, UT
Geographic Scope *See* Figure 2-1		West	North West	North East	North East	South	Central	South East	South West
Extreme Weather Pattern Number of Severe Weather Events reported between 01/01/2000 - 06/30/2010 *	county	Orange	Multnomah	Anne Arundel	Queens	Harris	Cook	Virginia Beach	Salt Lake
	Heavy rain	12	19	20	12	1	8	4	2
	Fog	13	0	14	0	0	0	0	4
	Hail	5	5	36	13	209	228	23	16
	Snow and ice	1	0	55	25	9	26	0	108
	Hurricane and tropical storm	0	0	1	2	6	0	0	0
Average Weather Pattern Monthly Surface Data Chart (*10-year average, 2000-2009*)** *See* Figure 2-2 through Figure 2-7		<ul><li>Number of days with greater than or equal to 0.1 inch precipitation</li><li>Number of days with greater than or equal to 0.5 inch precipitation</li><li>Number of days with greater than or equal to 1.0 inch precipitation</li><li>Number of days with snow depth greater than or equal to 1.0 inch</li><li>Total monthly precipitation (Hundredths of inches)</li><li>Total monthly snowfall (Tenths of inches)</li></ul>							

* Source : NOAA National Climatic Data Center (NCDC), U.S. Local Storm Event Database (http://www.ncdc.noaa.gov/oa/climate/severeweather/extremes.html)
** Source : NOAA National Climatic Data Center (NCDC), NNDC, Climate Data Online (http://www.ncdc.noaa.gov/oa/ncdc.html)

For each site, the extreme weather pattern is identified based on the number of severe weather events reported during the past 10 years from the U.S. Local Storm Event Database of the NOAA National Climatic Data Center (NCDC). NCDC receives Storm Data from the National Weather Service. The National Weather service receives their information from a variety of sources, which include but are not limited to: county, state and federal emergency management officials, local law enforcement officials and so on. Heavy rain, fog and snow here are different from what frequently occurs around the country on a regular basis because storm data documented in this database typically indicate significant weather phenomena having sufficient intensity to cause loss of life, injuries, significant property damage, and/or disruption to commerce. Some noticeable facts are: 1) Heavy rain is reported in Portland and Baltimore most frequently; 2) Baltimore experiences various extreme weather conditions quite evenly; 3) Houston and Chicago show a very strong tendency to have hail events; and 4) Salt Lake City shows high frequency of occurrence of snow events.

For the average weather pattern, monthly surface data are collected from NOAA National Climatic Data Center (NCDC) and averaged over 10 years from 2000 to 2009. Six measurements listed in Table 2-2 are plotted in Figure 2-2 through Figure 2-7. Figure 2-2, Figure 2-3 and Figure 2-4 indicate the amount of monthly rain precipitation for each site. These data do not tell us the exact intensity of rain events since the number of days is counted based on the total daily precipitation (e.g., $\geq$ 1 inch/day), not on the hourly precipitation intensity (inch/hr). However, roughly speaking, it can be assumed that Figure 2-4 indicates the number of days with relatively heavy rain while Figure 2-2 and Figure 2-3 represent the number of days with relatively light and moderate rain, respectively. Based on Figure 2-2, Figure 2-3, Figure 2-4 and Figure 2-6, Portland shows the most frequent occurrence of "light rain" while Virginia Beach, New York and Houston show the most frequent occurrence of "heavy rain". Irvine and Salt Lake City appear to be cities with the least precipitation. For the snow event, Figure 2-5 represents the number of days with a large amount of snow ($\geq$ 1 inch/day) and Figure 2-7 represents the total amount of snow for each month. Assuming that Figure 2-5 reflects frequency of observation of the heavy snow event, Salt Lake City and Chicago are cities with the heaviest snow events followed by Baltimore and New York.

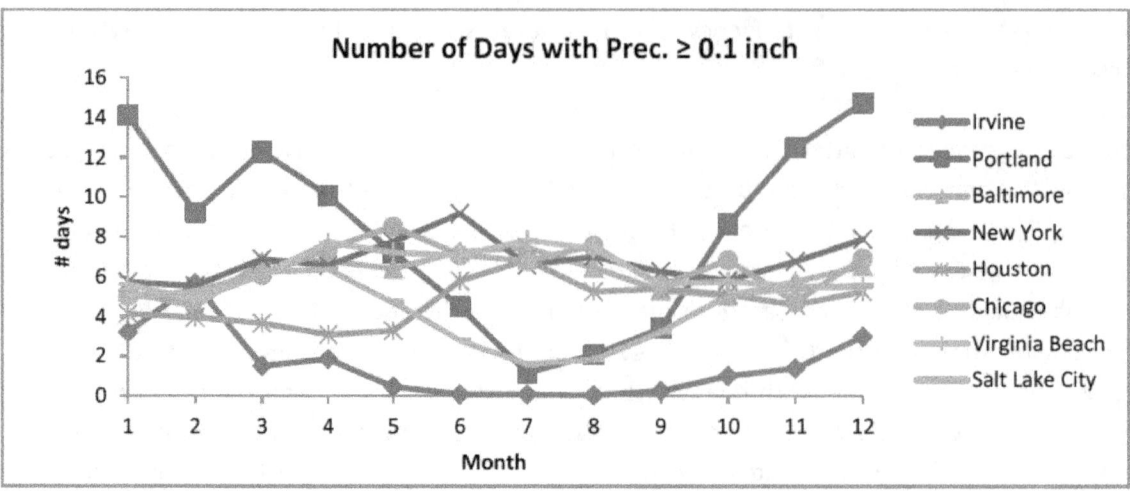

Figure 2-2. Monthly Surface Data: Number of Days with Rain Prec. ≥ 0.1 inch
(*10-year average, 2000-2009, Source: NOAA*)

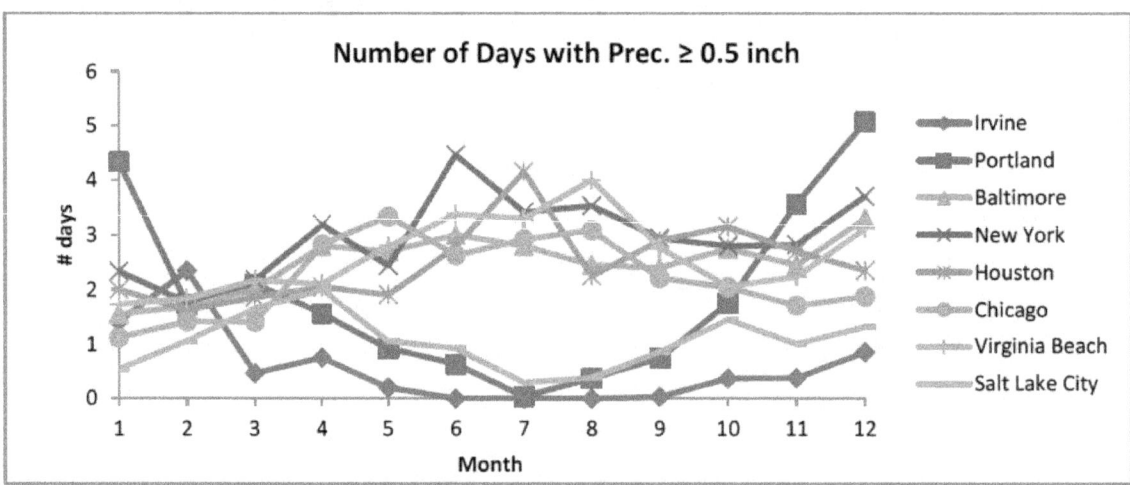

Figure 2-3. Monthly Surface Data: Number of Days with Rain Prec. ≥ 0.5 inch
(*10-year average, 2000-2009, Source: NOAA*)

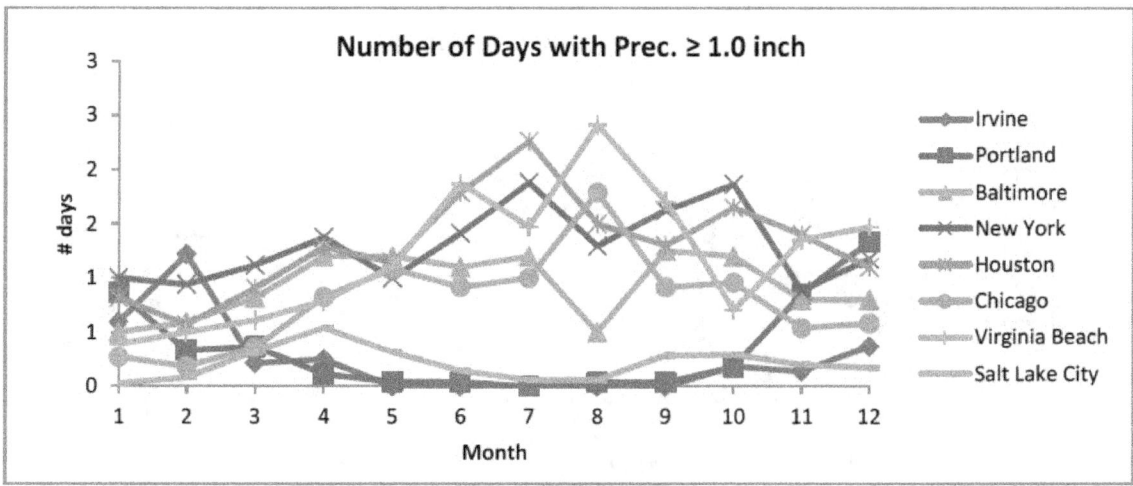

Figure 2-4. Monthly Surface Data: Number of Days with Rain Prec. ≥ 1.0 inch
(*10-year average, 2000-2009, Source: NOAA*)

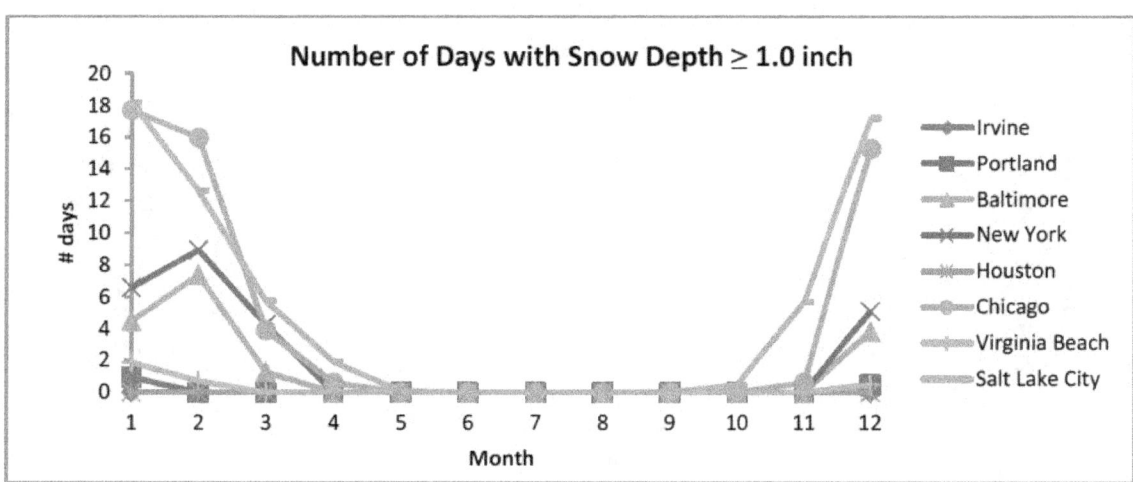

Figure 2-5. Monthly Surface Data: Number of Days with Snow Depth ≥ 0.1 inch
(*10-year average, 2000-2009, Source: NOAA*)

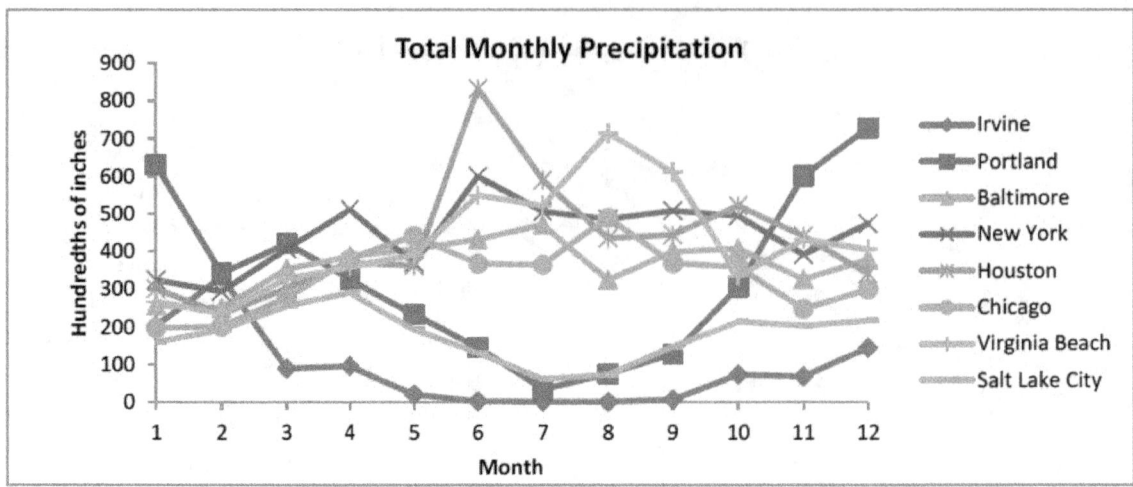

Figure 2-6. Monthly Surface Data: Total Monthly Rain Precipitation
(*10-year average, 2000-2009, Source: NOAA*)

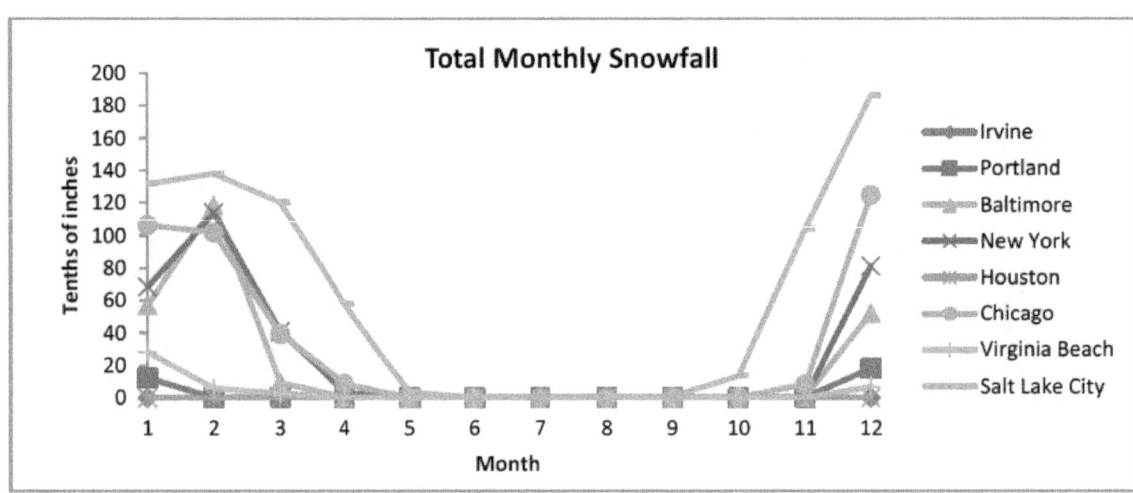

Figure 2-7. Monthly Surface Data: Total Monthly Rain Precipitation
(*10-year average, 2000-2009, Source: NOAA*)

2.1.3 Selection Results

Based on all information discussed in the previous section, this section provides a detailed recommendation for selection for candidate sites. To be able to select networks in a more systematic way, we try to quantify the characteristics of networks for each criterion using a ranking scale as presented in Table 2-3. For the criteria, the aforementioned issues related to traffic/weather data and the weather pattern are summarized and additional important criteria are also considered, most significant amongst those is the availability of a calibrated network model representation for simulation-based DTA, and consequently the associated effort envisioned to build, calibrate and validate such a model.

To avoid confusion in the interpretation of scale values, criteria are categorized into three types of questions related to Quality, Quantity and Agreement, and measured with the corresponding rating scale provided at the end of Table 2-3. Questions are described such that a larger value implies better suitability for selection in order for the sum of all the values for each network to represent the overall rating of each network. A weight is applied for each criterion, reflecting its relative importance. The default weight is 1. This approach is only used as a way to summarize in a consistent manner the relative desirability of each candidate network.

2.1.3.1 Weighted Criteria (Discussion for Table 2-3)

For traffic data, the first two items measure the overall availability of data that is identified in Table 2-1. When it comes to the traffic data for arterials, we have few detectors on arterials for a certain period for the Irvine network, but typically, traffic detectors are only on freeways for almost every network. From a practical perspective for the calibration process, the most critical factors are the quality of traffic data and the availability of the weather data from the weather station sufficiently close to traffic detectors. To reflect this concern, the weight of 2 is assigned to the corresponding two criteria: the quality of traffic data and the coverage of adjacent ASOS stations.

One of the most important factors for the weather-sensitive TrEPS model to be fully validated is the availability of various types and intensities of weather conditions. It has been recognized from the previous work that obtaining data for severe weather conditions is very difficult, especially with a sufficient amount of observations for calibration. Therefore we may need to assign our priority to a city with a high probability of occurrence of severe weather conditions. Two criteria associated with the weather pattern are included in Table 2-3; one for heavy rain and the other for snow. The weight values for both criteria are set to be 3.

Also critical to the ability of the project team to complete the objectives of the project successfully within the available resources is the availability of a network model of the area, and its readiness for simulation-based analysis. Accordingly, a weight of 3 has also been assigned to this criterion. Virginia Beach in particular suffers in this regard because of the absence of a previously developed network model that can readily support simulation analysis.

Table 2-3. Network Selection Criteria and Evaluation Results

Criteria [Measurement Type*]		Weight**	Irvine, CA	Portland, OR	Baltimore, MD	New York, NY	Houston, TX	Chicago, IL	Virginia Beach, VA	Salt Lake City, UT
Traffic Data	Are traffic data for freeway available? [QN]	1	5	5	5	4	4	5	5	5
	Are traffic data for arterials available? [QN]	1	2	1	1	2	1	1	1	1
	Quality of traffic detector data [QL]	2	4	2	2	2	1	3	3	4
	Are other data sources available?(e.g., mobile sensing data) [A]	1				Pending SHRP 2-L04 project	AVI tags	NAVTEQ		
Weather Data	Do weather stations (e.g., ASOS) cover the network? [QN]	2	5	5	3	4	4	5	4	3
	Are other data sources available? (e.g., Clarus system, local weather stations) [QN]	1	1	3	4	3	1	2	2	3
Weather Pattern	Probability of heavy rain (high rain intensity) [QN]	3	2	4	3	4	5	3	5	1
	Probability of snow event (high snow intensity) [QN]	3	1	2	3	3	1	4	2	5
Other Criteria	Are various types of highway/facilities included? (e.g., HOV, on/off-ramp) [A]	1	5	5	5	5	5	5	5	5
	Is the network ready to use? (No additional effort required to prepare the network) [QL]	3	5	2	3	4	3	2	1	3
	Is the size of the network sufficient to represent a major city in the U.S.? [A]	1	3	5	5	5	5	5	4	5
	Does local agency show interest? [QN] ***	1								
	Is local agency ready to apply weather-sensitive TrEPS models? [QN] ***	1								
	Total		33	34	34	36	30	35	32	35
	Weighted Total		55	52	52	59	48	61	55	60

* Criteria/questions are categorized into three measurement types and ranking scales for each type is as follows:

Scale	Measurement Type		
	Quality (QL)	Quantity (QN)	Agreement (A)
5	Excellent	Very much/Very high	Strongly Agree/Yes
4	Very Good	↑	Agree
3	Good	:	Neutral/Undecided
2	Fair	↓	Disagree
1	Poor	Very little/Very low	Strongly Disagree/No

** Different weights can be assigned criteria depending on the importance of each criterion.

*** Information regarding local agencies' interest and readiness was not incorporated in this initial evaluation.

2.1.3.2 Evaluation Results

The status of each network is ascertained with respect to each of the listed criteria. The results are presented in Table 2-3. Based on items for which evaluation for all eight networks is completed, the total scores and the weighted total scores are calculated and the corresponding ranking is presented in Figure 2-8. The four cities with the highest weighted totals are as follows:

1. **Chicago, IL**
2. **Salt Lake City, UT**
3. **New York, NY**
4. **Irvine, CA**

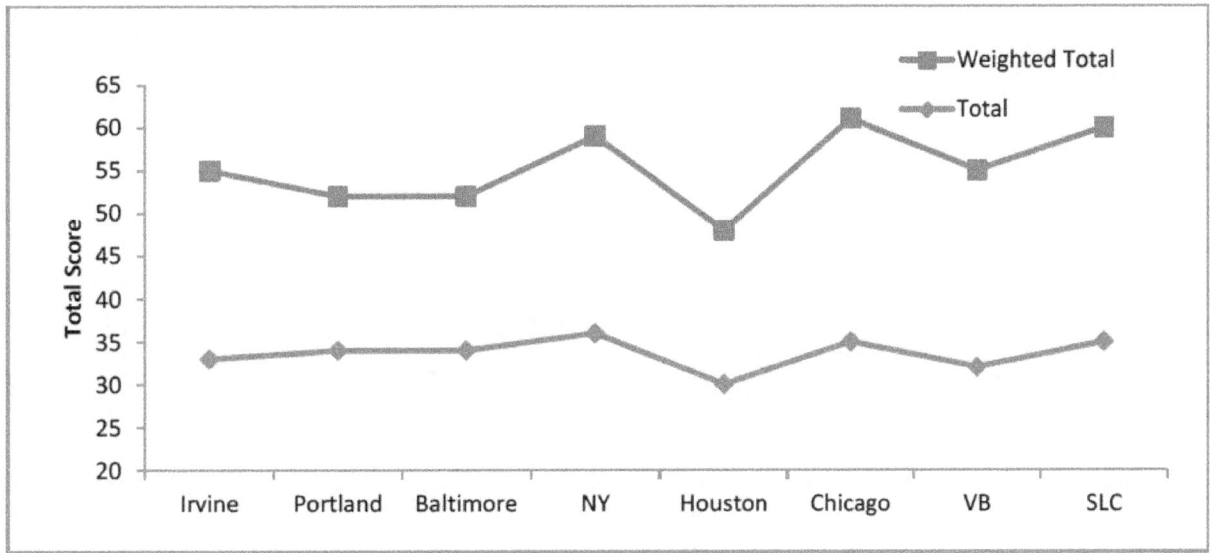

Figure 2-8. Network Rating

These four networks (i.e., Chicago, Salt Lake City, New York and Irvine) are selected for the calibration and validation for the TrEPS model.

2.2 Network Preparation

Simulation-based Dynamic Traffic Assignment (DTA) models require detailed network information. Networks used in DYNASMART are typically built on the basis of existing static networks, which often do not contain necessary information such as cycle and green times and allowed movements at each phase at a signalized intersection, or definition of each movement at a node (e.g. left turn, right turn, U- turn, and through movement). Thus, in addition to data provided by static networks, information from several other external sources is necessary to achieve an accurate representation of the real-world network. Figure 2-9 illustrates the overall process for building and converting networks for DYNASMART.

The main tool for this conversion is software called DYNABUILDER, which is capable of converting many networks from different platforms into a DYNASMART-P network. As DYNABUILDER also requires input files in a certain format, the pre-processing steps are conducted using several codes and macros to format the GIS or other sources of data.

The network preparation for four study sites was completed and snapshots for these networks are presented in Figure 2-10 through Figure 2-13. The figures provide the final network configuration and descriptive information on network components such as links, nodes and zones for all four networks, i.e., the New York (Long Island), Chicago, Salt Lake City and Irvine networks.

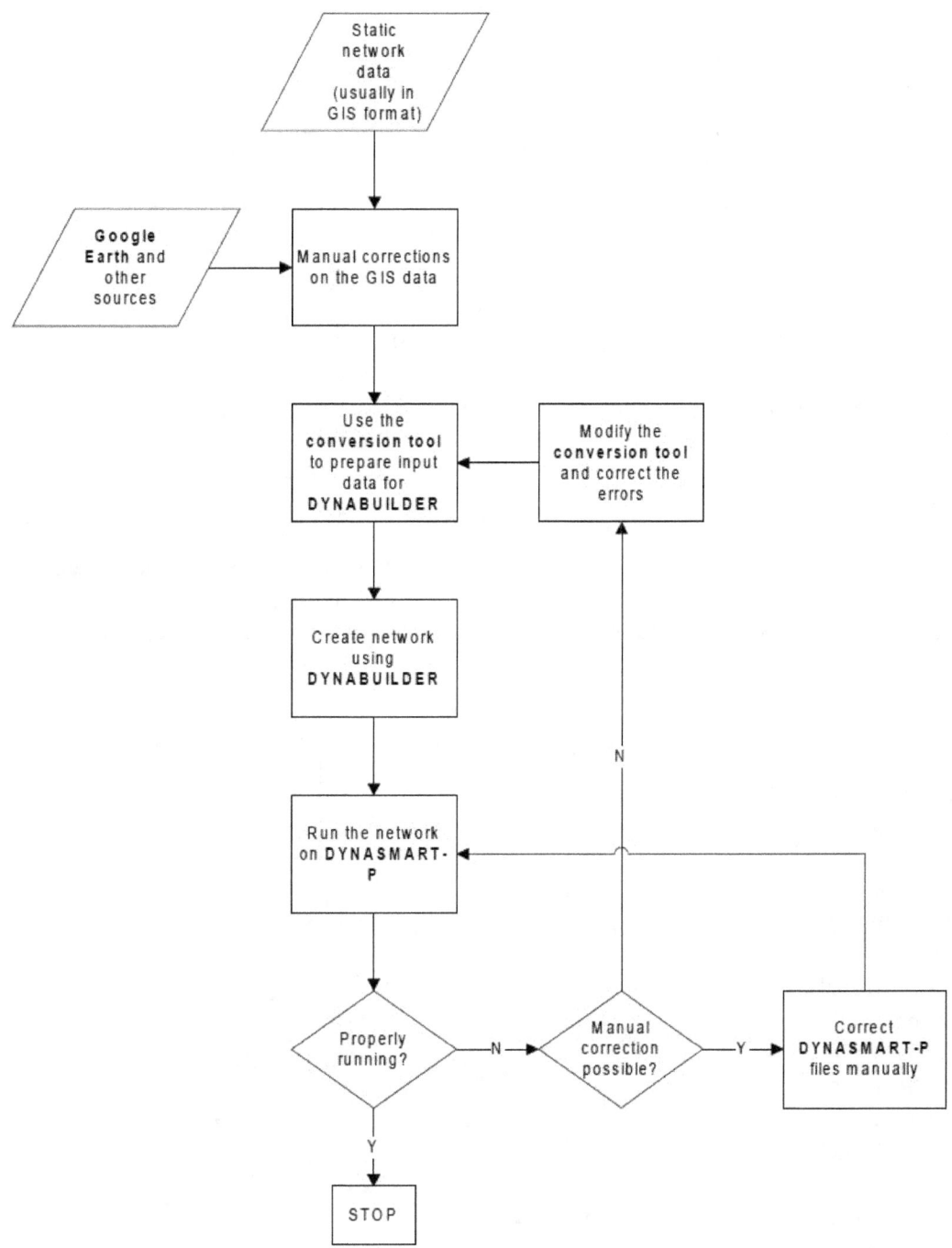

Figure 2-9. Flowchart for the Conversion from the Static to the Dynamic Network Model

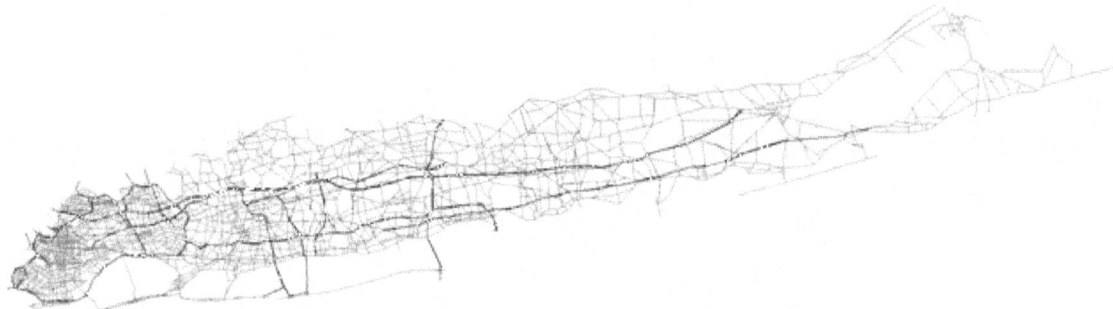

Network Description
- 21,791 links,
  - 1,588 freeways,
  - 14 links with tolls,
  - 31 highways,
  - 139 HOV facilities,
  - 2,087 ramps,
  - 17,945 arterials,
- 9,402 nodes,
  - 1,722 signalized intersections,
- 1,431 zones,
  - 1,421 internal,
  - 10 external,
- Demand period,
  - 6 am - 10 am,
  - 106 links with observations used in calibration.

Figure 2-10. Network Configuration and Description for New York Network (Long Island)

Network Description
- 40443 links
  - 144 links are tolled,
  - 1400 freeways
  - 201 highways,
  - 2120 ramps
  - (96 of them are metered),
  - 36722 arterials,
- 13,093 nodes,
  - 2,093 signalized intersections,
- 1961 zones,
  - 1,944 internal,
  - 17 external,
- Demand period,
  - 5 am - 10 am,
  - 355 links with observations used in calibration.

Figure 2-11. Network Configuration and Description for Chicago Network

Network Description
- 17,947 links
  - 791 freeways,
  - 136 highways,
  - 151 HOV facilities,
  - 576 ramps,
  - 16,293 arterials,
- 8,309 nodes,
  - 1,023 signalized intersections,
- 2,250 zones,
- Demand period,
  - 6 am - 9 am,
  - 66 links with observations are used in calibration.

Figure 2-12. Network Configuration and Description for Salt Lake City Network

Network Description
- 626 links
  - 91 freeways,
  - 99 ramps,
  - 436 arterials,
- 326 nodes,
  - 70
  - signalized intersections,
- 61 zones,
  - 41 internal,
  - 20 external,
- Demand period,
  - 4 am - 10 am,
  - 9 links with observations are used in calibration.

Figure 2-13. Network Configuration and Description for Irvine Network

2.3 Data Collection

The calibration of weather-sensitive TrEPS models requires two types of data: traffic and weather data. This section describes sources of the data for each study site and a detailed data collection procedure.

2.3.1 Weather Data

Weather data are available from two sources; the Automated Surface Observing System (ASOS) stations located at airports and the roadside Environmental Sensor Stations (ESS) available on the *Clarus* website (http://www.clarus-system.com). As the historical weather data from ESS have a time resolution of 20 minutes and are only available from 2009, ASOS data with the 5 minute resolution will be used in conjunction with traffic detector data collected and aggregated over a 5-minute interval. ASOS 5-minute weather data are available on the NOAA National Climatic Data Center (NCDC) website (ftp://ftp.ncdc.noaa.gov/pub/data/asos-fivemin). Table 2-4through Table 2-7 list airports in which ASOS stations are located for four study sites, respectively (i.e., New York, Chicago, Salt Lake City and Irvine), and time periods for which 5-min ASOS data are available from the above-mentioned website. Figure 2-14through Figure 2-17present the spatial distribution of ASOS stations for each respective network on the corresponding Google map, where circled areas represent a 10-mile radius of each station. It is noted that, however, for the NY network both traffic and weather data for the purpose of calibrating the weather-sensitive traffic flow model parameters are obtained from the nearby greater Baltimore area due to unavailability of comprehensive data sources for all desired items from the Long Island area. This is discussed in a more detail later in this section.

Table 2-4 Airports with ASOS Stations and Available Time Periods for Data (Long Island, NY)

No	Airport	Location	ICAO code	ASOS data
1	La Guardia Airport	Queens, NY	KLGA	2000 - present
2	John F. Kennedy International Airport	Queens, NY	KJFK	2000 - present
3	Republic Airport	Farmingdale, NY	KFRG	2005 - present
4	Long Island MacArthur Airport	Islip, NY	KISP	2000 - present
5	Brookhaven Airport	Shirley, NY	KHWV	2005 - present
6	Francis S. Gabreski Airport	Westhampton Beach, NY	KFOK	2005 - present

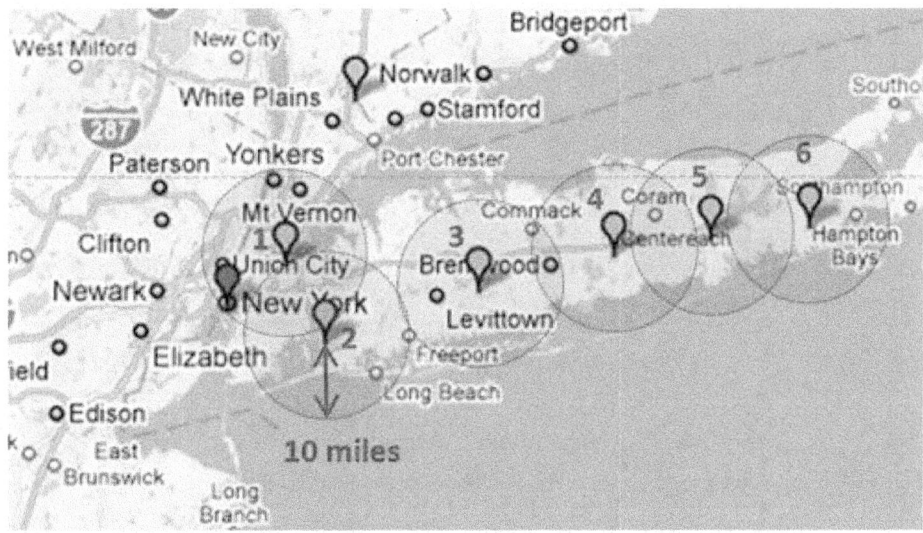

Figure 2-14 Long Island Study Area and Adjacent ASOS Stations
(Source: FAA, Surface Weather Observation Stations)

Table 2-5 Airports with ASOS Stations and Available Time Periods for Data (Chicago)

No	Airport	Location	ICAO code	ASOS data
1	Midway International Airport	Chicago, IL	KMDW	2005 - present
2	O'Hare International Airport	Chicago, IL	KORD	2000 - present
3	Dupage County Airport	Dupage, IL	KDPA	2005 - present
4	Chicago Executive Airport	Cook, IL	KPWK	2005 - present
5	Aurora Municipal Airport	Kane, IL	KARR	2005 - present

Figure 2-15. Chicago Study Area and Adjacent ASOS Stations
(Source: FAA, Surface Weather Observation Stations)

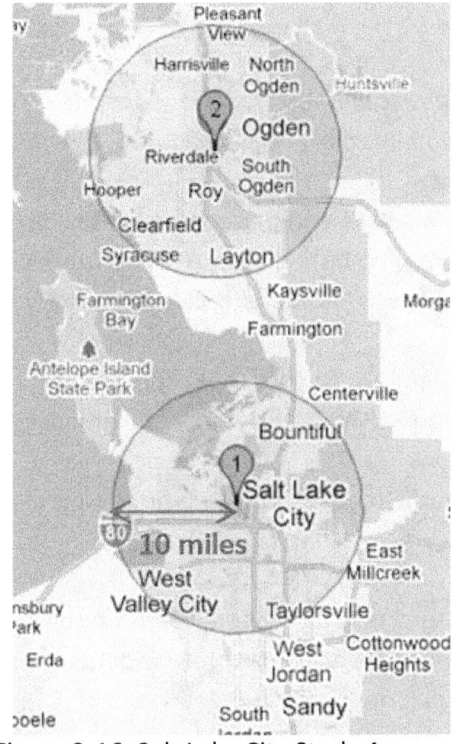

Figure 2-16. Salt Lake City Study Area and
Adjacent ASOS Stations
(Source: FAA, Surface Weather Observation Stations)

Table 2-6. Airports with ASOS Stations and Available Time Periods for Data (SLC)

No	Airport	Location	ICAO code	ASOS data
1	Salt Lake City International Airport	Salt Lake City, UT	KSLC	2000 - present
2	Ogden-Hinckley Airport	Weber, UT	KOGD	2005 - present

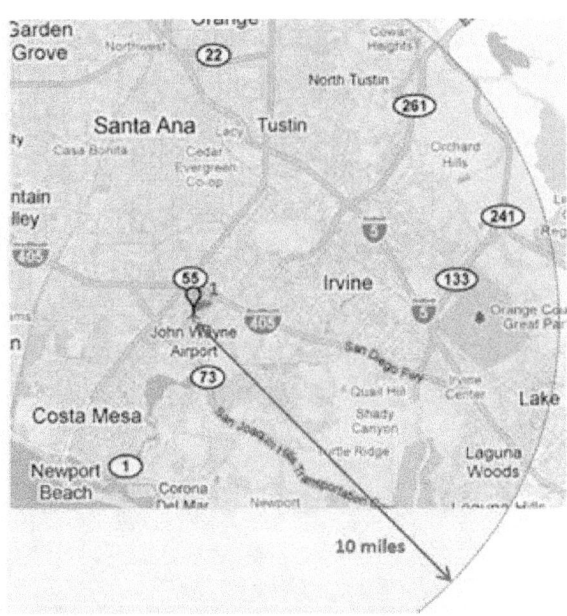

Figure 2-17. Irvine Study Area and
Adjacent ASOS Stations
(Source: FAA, Surface Weather Observation Stations)

Table 2-7. Airports with ASOS Stations and Available Time Periods for Data (Irvine)

No	Airport	Location	ICAO code	ASOS data
1	John Wayne Airport	Santa Ana, CA	KSNA	2005 – present

2.3.2 Traffic Data

The primary source of traffic data for supply-side parameter calibration is loop detectors installed on freeway lanes. Historical data with the 5-minute aggregation interval are used and the time periods for the data vary with the study site over the 2005-2011 period.

In selecting detector locations and collecting the data, the following criteria are mainly considered.

1. Choose detectors as close as possible to ASOS stations; no farther than 10 miles from ASOS.
2. Remove the influence of other external events such as incidents/accidents, work zones and planned special events.
3. Include various facility/lane types and calibrate separately for each type. For instance, types can be classified into mainlines, on-ramps, off-ramps and HOV; and the number of lanes could be further distinguished.
4. Find segments that experience a wide range of traffic regimes, i.e., free-flow, stop-and-go and congested states.

Note that the process for removing the effect of external events is highly dependent on the availability of other event data. In case where there is difficulty obtaining detailed data for incidents, work zones, and special events, one could focus on traffic data and clean outliers in the dataset only. Since we are averaging measures over a long period of time, at least one year, the influence of other external events on traffic parameters is expected to be very small.

2.3.2.1 Chicago

For the Chicago network, traffic data are obtained from Illinois DOT. 5-minute aggregated data from 2009 are used. Figure 2-18shows a map of the selected detector locations in Chicago. At each location, traffic data from north- or south-bound directions are obtained. There is no HOV lane at any of the selected locations.

Figure 2-18. Selected Detector Locations in Chicago
(Source: Google Map, Accessed April, 2011)

2.3.2.2 Salt Lake City

For the Salt Lake City network, traffic data are obtained from Utah Freeway Performance Measurement System (PeMS). 5-minute aggregated data from 2009 are used. Figure 2-19 shows a map of the selected detector locations in Salt Lake City. At each location, traffic data from north- and south-bound directions are obtained. Two out of ten of selected locations include an HOV lane at both directions. Traffic data from HOV lanes are also obtained.

Figure 2-19. Selected Detector Locations in Salt Lake City
(Source: Google Map, Accessed April, 2011)

2.3.2.3 Irvine

For the Irvine network, traffic data are obtained from California Freeway Performance Measurement System (PeMS). Five-minute aggregated data from 2005-2007 are used. Figure 2-20 shows a map of the selected detector locations in Irvine. At each location, traffic data from north- and south-bound directions are obtained. Three out of five selected locations include an HOV lane at both directions. Traffic data from HOV lanes are also obtained.

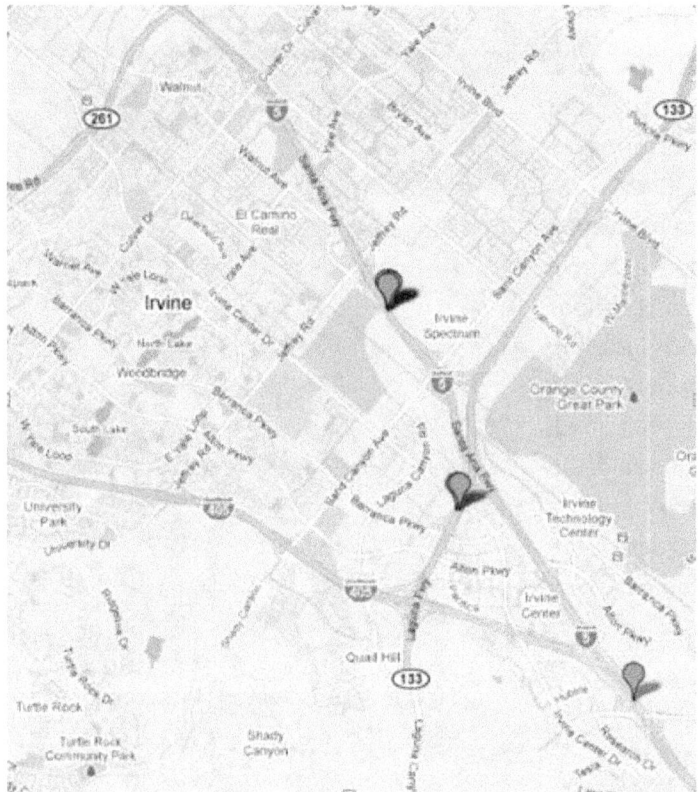

Figure 2-20. Selected Detector Locations in Irvine (Source: Google Map, Accessed April, 2011)

2.3.2.4 Baltimore (substitution for New York)

Despite substantial efforts to obtain traffic data from the Long Island area, we were not able to identify comprehensive data source for all desired items. Fortunately, the items missing are not unique to the Long Island region, as they pertain to the traffic flow aspects under certain weather adjustment factor (WAF) across similar areas. Accordingly, to advance the progress in estimating model parameters for the New York network, other sources of traffic data have been investigated focusing on adjacent states such as New Jersey, Pennsylvania, and Maryland, Based on the data availability and the general characteristics (e.g., social/geographical characteristics and weather pattern), data from the Baltimore area were retained for this purpose. We believe that these data can be a good representative of New York data as Baltimore is a large metropolitan area with a similar geography (i.e., located on the northeast coast). Furthermore the I-95 Corridor through Baltimore and Maryland is heavily traveled by drivers from New York and New Jersey.

The traffic data are collected from loop detectors installed on freeways along I-695 and the time period covers 2010 and 2011. Locations of the weather station and selected detectors are presented in Figure 2-21.

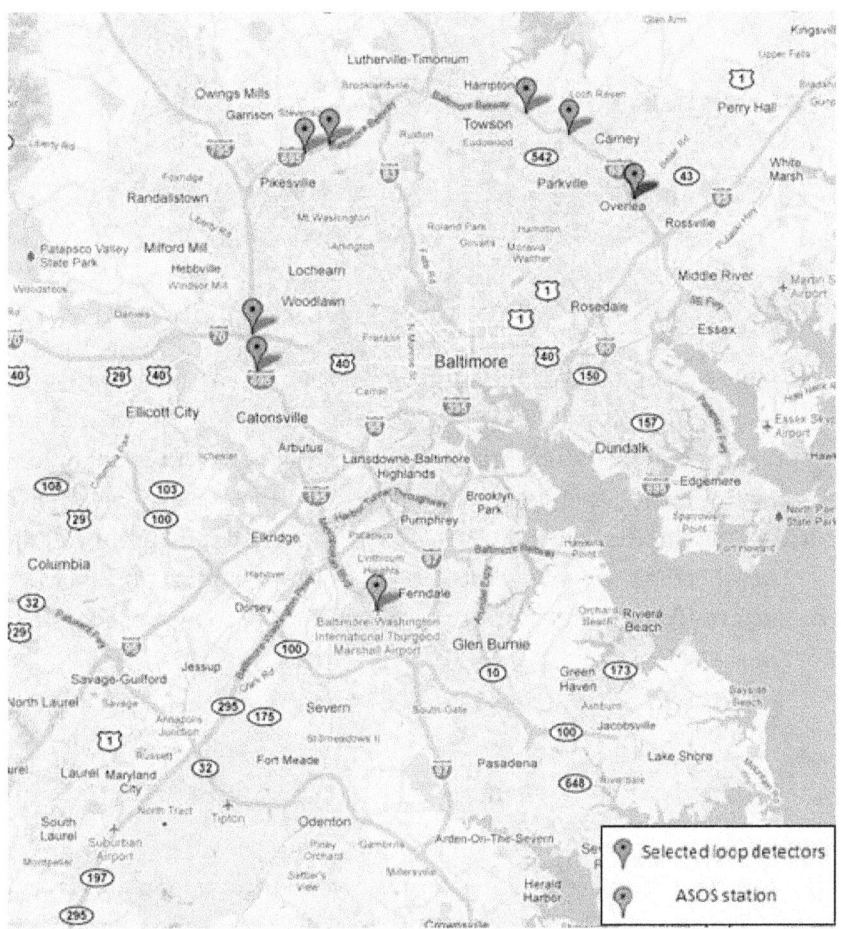

Figure 2-21. Locations for Selected Detectors and ASOS Station in Baltimore (Source: Google Map, Accessed September, 2011)

3. Calibration and Validation of Weather-sensitive TrEPS Model

3.1 Supply-side Parameter Calibration

The supply-side parameter calibration for this study includes two parts: calibrating parameters in the traffic flow model (i.e., modified Greenshields models) and estimating the weather adjustment factor (WAF). First, the traffic flow model is calibrated under different weather conditions based on pre-defined weather categories. The calibrated parameters for the normal weather are supplied to DYNASMART as the base case traffic flow model. The parameters under different weather conditions are used to obtain the weather adjustment factor (WAF), which is a reduction factor that reflects the weather impact on each traffic flow parameter. The detailed calibration procedure and the results are discussed in the following sections.

3.1.1 Calibration of Traffic Flow Model Parameters

3.1.1.1 Data Preparation

Traffic data used for the calibration are three major observations from loop detectors, i.e., link volume (or flow rates), occupancy and speed. All traffic data have the aggregation interval of 5 minutes. The occupancy data are further converted into the density using the following relationship *(Cassidy and Coifman, 1997)*:

$$k = \frac{52.8}{L_v + L_s} \cdot occ \qquad (3\text{-}1)$$

where

k	=	density [veh/mi/lane]
L_v	=	average vehicle length [feet]
L_s	=	average sensor length [feet]
occ	=	occupancy [%]

L_v is assumed to be 5 meters (approximately 16.4 feet); and L_s is set to 2 meters (approximately 6.5 feet). Weather data are collected from nearby Automated Surface Observing System (ASOS) stations located at airports, which contain 5-minute aggregated information of visibility, rain intensity level and snow intensity level. Traffic data and weather data are then matched together according to the timestamps to classify each traffic observation into different weather categories.

3.1.1.2 Modified Greenshields Traffic Flow Model

Two types of modified Greenshields models are used in DYNASMART for traffic propagation *(Mahmassani and Sbayti, 2009)*. Type 1 is a dual-regime model in which constant free-flow speed is specified for the free-flow conditions (1st regime) and a modified Greenshields model is specified for congested-flow conditions (2nd regime) as shown in Figure 3-1.

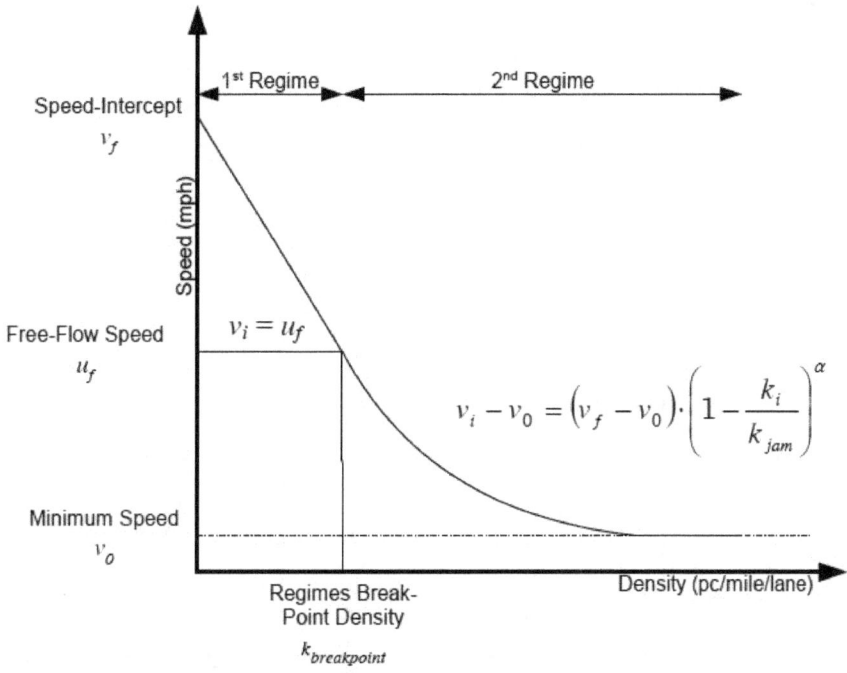

Figure 3-1. Type 1 modified Greenshields model (dual-regime model)
(Source: DYNASMART-P User's Guide, Accessed June, 2011)

In mathematical terms, the Type 1 modified Greenshields is expressed as follows:

$$v_i = \begin{cases} u_f & 0 < k_i < k_{breakpoint} \\ v_0 + (v_f - v_0)\left(1 - \dfrac{k_i}{k_{jam}}\right)^\alpha & k_{breakpoint} < k_i < k_{jam} \end{cases} \quad (3\text{-}2)$$

where v_i = speed on link *i*

v_f = speed-intercept

u_f = free-flow speed on link *i*

v_0 = minimum speed on link *i*

k_i = density on link i

k_{jam} = jam density on link i

α = power term

$k_{breakpoint}$ = breakpoint density

Type 2 uses a single-regime to model traffic relations for both free- and congested-flow conditions as shown in Figure 3-2.

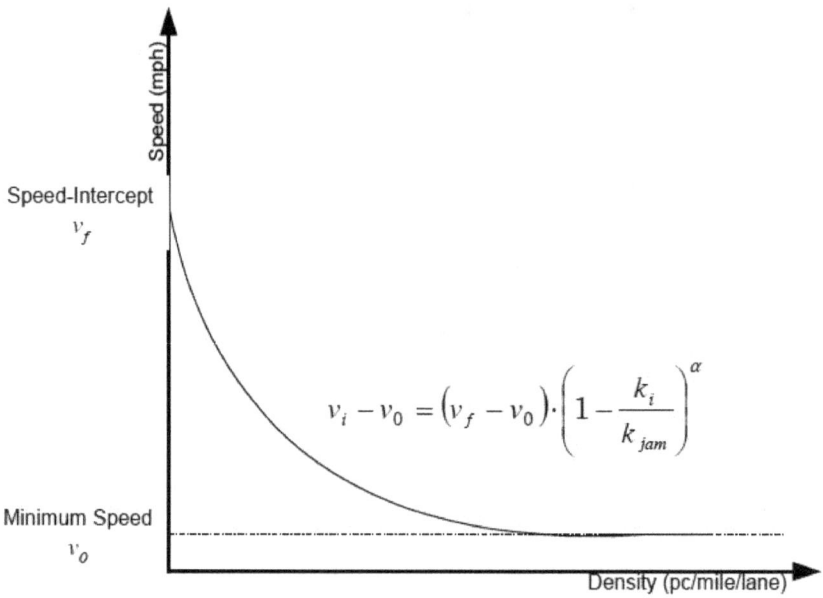

Figure 3-2. Type 2 modified Greenshields model (single-regime model)
(Source: DYNASMART-P User's Guide, Accessed June, 2011)

In mathematical terms, the type 2 modified Greenshields is expressed as follows:

$$v_i - v_0 = (v_f - v_0) \cdot \left(1 - \frac{k_i}{k_{jam}}\right)^{\alpha} \qquad (3\text{-}3)$$

Dual-regime models are generally applicable to freeways, whereas single-regime models apply to arterials. The reason why a two-regime model is applicable for freeways in particular is that freeways have typically more capacity than arterials, and can accommodate dense traffic (up to 2300 pc/hr/ln) at near free-flow speeds. On the other hand, arterials have signalized intersections, meaning that such a phenomenon may be short-lived, if present at all. Hence, a slight increase in traffic would elicit more deterioration in prevailing speeds than in the case of freeways. Therefore, arterial traffic relations are better explained using a single-regime model. All the traffic data used in this study come from loop detectors installed on highways. Therefore the dual-regime model is chosen to fit the collected historical data. For the dual regime model, the total six parameters are calibrated, namely, breakpoint density (k_{bp}), free flow speed (u_f), speed-intercept (v_f), minimum speed (v_0), jam density (k_{jam}), and the shape parameter (α). For the single regime model, only three parameters including speed-intercept (v_f), minimum speed (v_0), and the shape parameter (α) are used.

3.1.1.3 Weather Categorization

The weather categories were defined based on the precipitation type and the intensity. With a normal weather as the base case, in which no precipitation is observed, three levels of precipitation intensities (light, moderate and heavy) are used for both rain and snow. Table 3-1 shows these seven weather categories and the corresponding precipitation intensity ranges: normal (no precipitation), light rain (intensity less than 0.1 in./hr), moderate rain (0.1 to 0.3 in./hr), heavy rain (greater than 0.3 in./hr), light snow (less than 0.05 in./hr), moderate snow (0.05 to 0.1 in./hr), and heavy snow (greater than 0.1 in./hr). The values for the intensity range are based on the literature *(Federal Meteorological Hand Book, 2005; Hranac et al., 2006; Maze et al., 2006)*.

The categories were further adjusted (i.e., merged or dropped) during the calibration process if there were not sufficient traffic observations for a certain weather category. This happened to the Irvine network for all snow-related categories. For the Salt Lake City and Chicago networks, the moderate and heavy categories were merged for both rain and snow since traffic data for heavy rain/snow were not sufficiently covering the whole density range to enable regressions to be carried out. A summary of weather categorization for different networks is given in Table 3-1.

Table 3-1. Weather categorization for different networks

Network	Weather Condition (precipitation intensity (inch/hr))						
	normal (no precip.)	light rain (< 0.1)	moderate rain (0.1 - 0.3)	heavy rain (> 0.3)	light snow (< 0.05*)	moderate snow (0.05 - 0.1*)	heavy snow (> 0.1*)
Irvine	✓	✓	✓	✓			
Salt Lake City	✓	✓	✓		✓	✓	
Chicago	✓	✓	✓		✓	✓	
Baltimore	✓	✓	✓	✓	✓	✓	✓

* Liquid Equivalent Snowfall Intensity

3.1.1.4 Calibration Procedure and Results

After traffic data are categorized, parameters in the modified Greenshields model are calibrated for each weather condition using a nonlinear regression approach. The following steps describe the procedures for calibrating the dual-regime model, which is used in most cases when traffic data are collected from freeways.

Step 1. Plot the speed vs. density graph, and set initial values for all the parameters, i.e. breakpoint density (k_{bp}), speed-intercept (v_f), minimum speed (v_0), jam density (k_{jam}), and the shape parameter (α), based on observations.

Step 2. For each observed density (k_i), calculate the predicted speed value ($\hat{v}_i$) using Eq. (2) and the parameters initialized in Step 1.

Step 3. Compute the squared difference between observed speed value (v_i) and predicted speed value ($\hat{v}_i$), for each data point, and sum the squared error over the entire data set.

Step 4. Minimize the sum of squared error obtained in Step 3, by changing the values of model parameters.

Unlike the linear regression used in the previous research (*Mahmassani, et al., 2009*), which divides the data into two parts and estimates the two regimes separately, the nonlinear regression used in this study allows estimating the model as a whole, which gives a smooth joint point at the breakpoint density. Step 4 is implemented by Microsoft Excel Solver which uses the generalized reduced gradient algorithm to find the optimal solution. Based on the observed data, the minimum speed (v_0) and jam density (k_{jam}) turn out to be insensitive to weather conditions. For Irvine network, the minimum speed is assumed to be 10 mph, while for Chicago and Salt Lake City, a minimum speed of 2 mph is used. The jam density is assumed to be 225 vpmpl for all the three networks.

The goodness-of-fit of the nonlinear regression model can be measured by the root mean square error (RMSE) as shown in Eq. (3-4), where $\hat{v}_i$ is the predicted/modeled value and v_i is the observed value for the ith observation in the sample with the size of N. The smaller the RMSE is, the better the model represents the data.

$$RMSE = \sqrt{\frac{1}{N}\sum_{i=1}^{N}(v_i - \hat{v}_i)^2} \qquad (3-4)$$

Another measurement is the R-squared value, which is computed in the same way as in linear regression models. The expression is shown in Eq. (3-5), where $\bar{v}$ represents the mean of the observed data. The R-squared value is the ratio of the regression sum of squares to the total sum of squares, which explains the proportion of variance accounted for in the dependent variable by the model *(StatSoft, Inc., 2011)*. The closer R-squared value is to 1, the better the model fits the data.

$$R^2 = 1 - \frac{SSE}{SST} = 1 - \frac{\sum (v_i - \hat{v}_i)^2}{\sum (v_i - \bar{v})^2} \qquad (3-5)$$

Examples of calibrated speed-density curves for each network are presented in Figure 3-3. It is observed that the overall speed for both uncongested and congested regimes decreases as the weather conditions become severe. The snow event, especially the moderate and heavy snow, causes the clear reductions in speed as shown in Chicago, Salt Lake City and Baltimore networks. Detailed calibration results for all study sites are presented in **Appendix C**.

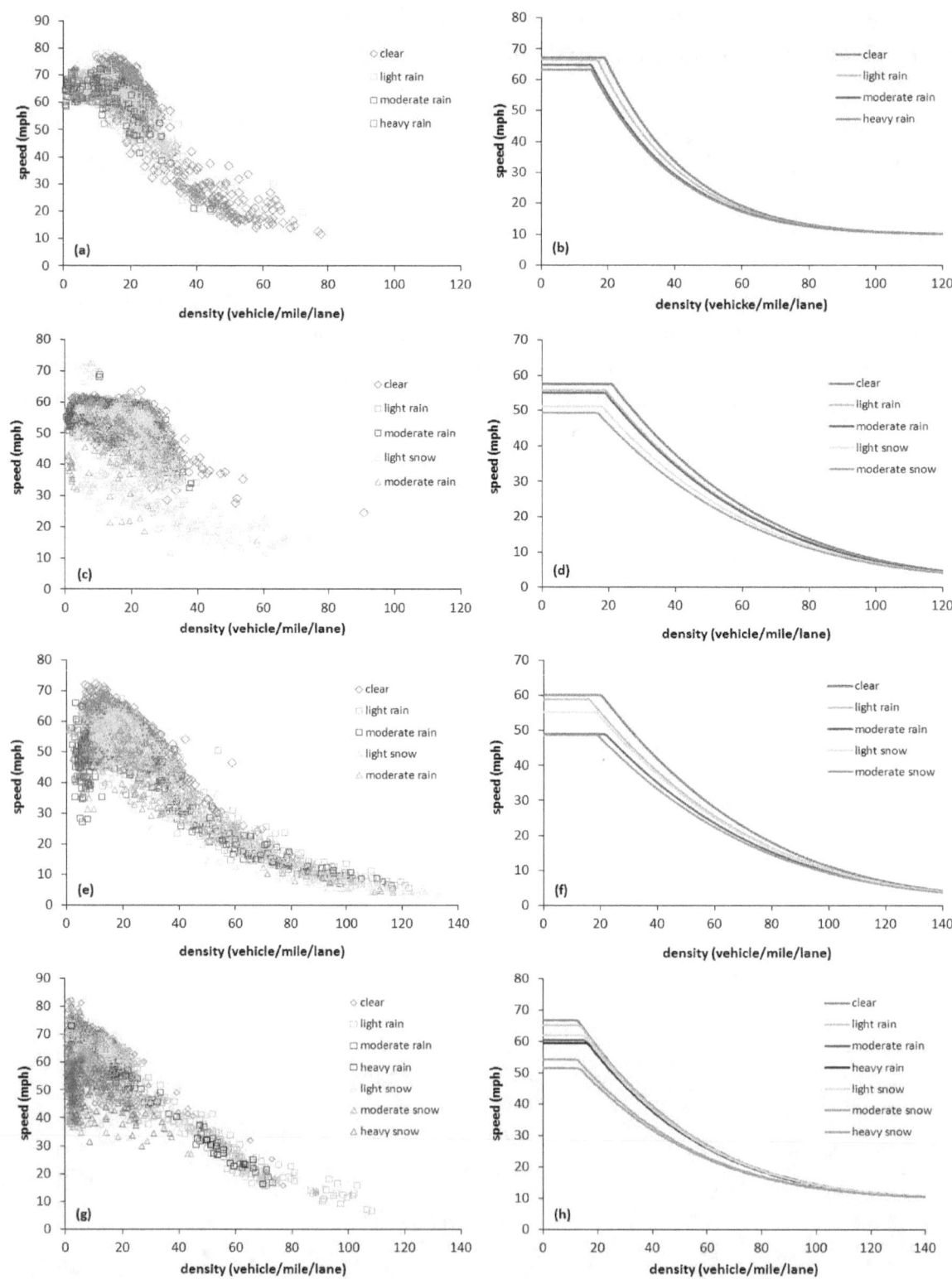

Figure 3-3. Examples of raw traffic data and calibrated speed-density curves under different weather conditions for each network: Irvine (a,b), Salt Lake City (c,d), Chicago (e,f) and Baltimore (g,h).

3.1.2 Calibration of Weather Adjustment Factor (WAF)

In DYNASMART, supply-side parameters that are expected to be affected by the weather condition are identified as presented in Table 3-2. The inclement weather impact on each of these parameters is represented by a corresponding weather adjustment factor (WAF) such that

$$f_i^{Weather\ Event} = F_i \cdot f_i^{Normal} \tag{3-6}$$

where $f_i^{Weather\ Event}$ denotes the value of parameter *i* under a certain weather event, f_i^{Normal} denotes the value of parameter *i* under the normal condition and F_i is the WAF for parameter *i*.

Table 3-2. Supply Side Properties related with Weather Impact in DYNASMART

Category	i	Parameter Description
Traffic flow model[1]	1	Speed-intercept (mph)[1]
	2	Minimal speed (mph)
	3	Density break point (pcpmpl)[1]
	4	Jam density (pcpmpl)
	5	Shape term alpha
Link performance	6	Maximum service flow rate (pcphpl or vphpl)
	7	Saturation flow rate (vphpl)
	8	Posted speed limit adjustment margin (mph)
Left-turn capacity	9	g/c ratio
2-way stop sign capacity	10	Saturation flow rate for left-turn vehicles(vphpl)
	11	Saturation flow rate for through vehicles(vphpl)
	12	Saturation flow rate for right-turn vehicles(vphpl)
4-way stop sign capacity	13	Discharge rate for left-turn vehicles(vphpl)
	14	Discharge rate for through vehicles(vphpl)
	15	Discharge rate for right-turn vehicles(vphpl)
Yield sign capacity	16	Saturation flow rate for left-turn vehicles(vphpl)
	17	Saturation flow rate for through vehicles(vphpl)
	18	Saturation flow rate for right-turn vehicles(vphpl)

1) only available in dual-regime model
Source: Mahmassani et al. ,2009

The WAF is assumed to be a linear function of weather conditions, and is expressed in the following form

$$F_i = \beta_{i0} + \beta_{i1} \cdot v + \beta_{i2} \cdot r + \beta_{i3} \cdot s + \beta_{i4} \cdot v \cdot r + \beta_{i5} \cdot v \cdot s \qquad (3\text{-}7)$$

where

F_i weather adjustment factor for parameter i,

v visibility (mile),

r precipitation intensity of rain (inch/hr),

s precipitation intensity of snow (inch/hr), and

$\beta_{i0}, \beta_{i1}, \beta_{i2}, \beta_{i3}, \beta_{i4}, \beta_{i5}$ coefficients to be estimated.

Thus, once the speed-density functions for different weather conditions (i.e., normal, light rain, moderate rain, etc.) are obtained for each network, a linear regression analysis is performed to obtain the WAF for each parameter based on observed rain intensities, snow intensities and visibility levels. A detailed description of the calibration procedure is provided below.

3.1.2.1 Calibration Procedure

The calibration of coefficients in Eq. (3-7) includes the following steps.

Step 1. For each weather condition c, calculate the WAF for each parameter i such that $F_i = f_i^c / f_i^{Base} \quad \forall c$, where *Base* denotes the normal (no precipitation) weather.

Step 2. Assign F_i to corresponding traffic-weather data such that each observation has a structure similar to the following:
{time, traffic data (volume, speed, density), weather data(v, r, s), WAF($F_1, \cdots, F_i$)}.

Step 3. For each parameter i, estimate coefficients $\beta_{i0}, \beta_{i1}, \beta_{i2}, \beta_{i3}, \beta_{i4}, \beta_{i5}$ by conducting the regression analysis using Eq. (3-7) given F_i as a dependent variable and weather data (v, r, s) for all observations as independent variables.

Note that not all of the parameters listed in Table 3-2 can be calibrated using the observation data. Some parameters could be inferred from other calibrated parameters.

(1) Traffic flow model related parameters, that is, speed-intercept (v_f), minimum speed(v_0), density break point(k_{bp}), jam density(k_{jam}), shape term alpha(α) and maximum service flow rate (q_{max}) can be calibrated from the traffic data. However, as minimum speed and jam density turn out to be insensitive to weather conditions from the calibration results, WAF for those parameters are assumed as 1, which indicates these are not affected by weather conditions. In addition, the shape parameter alpha is also fixed as 1 based on the observations that the both speed-intercept (v_f) and alpha(α) govern the shape of the curve and controlling for one variable results in a more consistent and meaningful pattern on the other allowing a better interpretation.

(2) Link characteristics: saturation flow rate, and posted speed limit adjustment could be inferred from the calibrated traffic flow model.

(3) Signal control: the adjustments in cycle length, offset, green, amber, maximum green, and minimum green could be inferred from the saturation flow rate.

(4) Left turn/stop sign/yield sign capacities could be calibrated using the traffic data, for example, maximum observed flow rate could be used as a surrogate of capacity.

3.1.2.2 Calibration Results

Based on the calibrated traffic model of the three networks, it is found that the maximum service flow rate (q_{max}), shape parameter (α), and free flow speed (u_f), are sensitive to both rain and snow intensities. As the rain or snow intensity increases, maximum flow rate, speed intercept and free flow speed are reduced. Similar findings are present in the literature *(Ibrahim and Hall, 1994; Rakha et al., 2008)*. It is also found that increasing snow intensity reduces breakpoint density; however, the effect of rain on it is not as clear as that of snow, as in some networks it decreases with rain intensity (e.g., Irvine) while in other cases it increases (e.g., Baltimore). As a summary, the effects of the rain intensity and the snow intensity on different traffic flow model parameters are presented in Figure 3-4 and Figure 3-5, respectively.

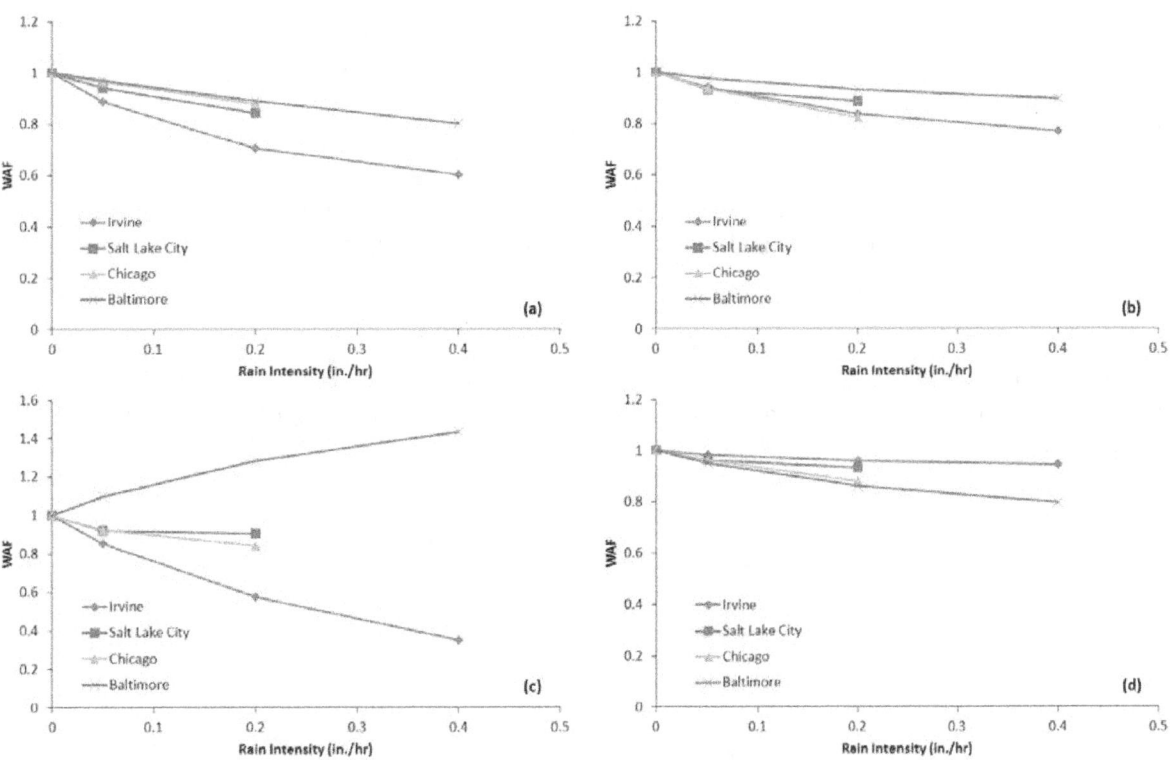

Figure 3-4. Effect of the rain intensity on weather adjustment factors for: (a) maximum flow rate (q_{max}); (b) speed intercept (v_f); (c) breakpoint density (k_{bp}); and (d) free flow speed (u_f)

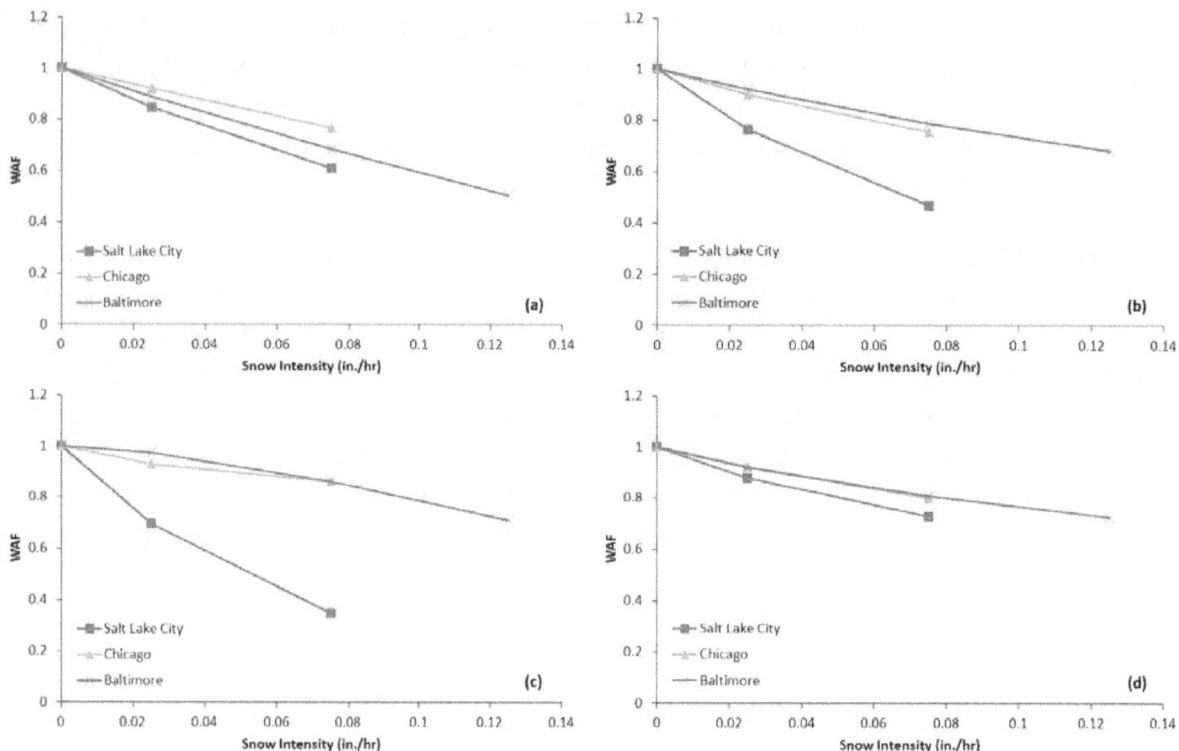

Figure 3-5. Effect of the snow intensity on weather adjustment factors for: (a) maximum flow rate (q_{max}); (b) speed intercept (v_f); (c) breakpoint density (k_{bp}); and (d) free flow speed (u_f)

The calibration results of WAF for the three networks are provided in Table 3-3. The low R-squared values of breakpoint density (k_{bp}) suggest that this parameter is insensitive to visibility and precipitation intensity levels.

Table 3-3. Calibration results of WAF

Network	Parameter	β_0	β_1	β_2	β_3	β_4	β_5	R^2
Irvine	q_{max}	0.8424	0.0154	0.0244	0	-0.1942	0	0.7251
	v_f	0.9188	0.0080	-0.0665	0	-0.0965	0	0.7227
	k_b	0.8203	0.0178	-0.5202	0	-0.2078	0	0.4305
	u_f	0.9778	0.0022	0.0033	0	-0.0268	0	0.3704
Salt Lake City	q_{max}	0.9202	0.0077	-0.1242	-2.8739	-0.0801	-0.3076	0.6361
	v_f	0.7887	0.0209	0.8547	-0.6376	-0.1641	-0.8786	0.8187
	k_b	0.6933	0.0305	1.4373	0.8021	-0.2161	-1.3046	0.4389
	u_f	0.8993	0.0098	0.4110	-0.6111	-0.0887	-0.4044	0.8748
Chicago	q_{max}	0.9979	0.0003	-0.3312	-3.0583	-0.0436	-0.0046	0.6919
	v_f	0.9254	0.0071	-0.1071	-1.6901	-0.1026	-0.1902	0.9061
	k_b	0.8713	0.0122	0.5052	0.1758	-0.1700	-0.2138	0.2413
	u_f	0.9702	0.0029	-0.2695	-1.8068	-0.0437	-0.1150	0.7569
Baltimore	q_{max}	0.9874	0.0015	-0.3753	-3.3884	-0.0243	-0.1267	0.6397
	v_f	0.9570	0.0044	-0.0738	-1.8262	-0.0294	-0.1302	0.6987
	k_b	1.0894	-0.0081	0.3924	-3.5266	0.1371	0.1888	0.2572
	u_f	0.9303	0.0068	-0.1044	-1.1713	-0.0733	-0.1662	0.8466

3.2 Demand-side Parameter Calibration

The demand-side parameter calibration for this study includes several considerations: the base-case OD matrix estimation, changes in dynamic OD trip patterns due to weather conditions, user responses to information and various advisory/control operations schemes, and so on. As an immediate task associated with the network building procedure, the base-case OD matrix estimation has been given priority in implementation plans for the demand-side parameter calibration. The base-case OD matrix here indicates a time-dependent OD matrix under the normal weather condition. As an effort to address the second consideration (demand pattern change under inclement weather conditions), however, the OD matrices for various adverse weather cases can also be estimated and investigated to find a certain type of factor that reflects a structural adjustment in the demand pattern. The present document only includes the estimation procedure and results for the base-case OD matrices for the study sites. Further improvement can be made while the on-line TrEPS is implemented in terms of capturing demand patterns under different weather conditions as the real-time traffic data fed into the TrEPS model reflect the traffic state influenced by the prevailing weather.

3.2.1 Estimating Base Case OD Matrix

3.2.1.1 Estimation Procedure

Time-dependent (or dynamic) origin-destination (TDOD) matrices are of crucial importance as an input for dynamic traffic assignment (DTA) models. Determining the scale and resolution of the network model is an essential step in planning applications, with important implications for specifying the associated time-dependent demand patterns. In order to capture the time-dependent pattern, a bi-level optimization method is used *(Verbas et al., 2001)*.

The inputs to this framework are:
- Static/historical OD matrix for the planning time horizon,
- Time-dependent traffic counts on selected observation links.

The output is:
- Time-dependent OD matrices over the time horizon with a chosen time interval (usually 5 or 15 minutes).

The resolution of the time intervals depends on the resolution of the link counts available. Although a higher resolution (smaller time intervals) are usually desirable, it must be noted that travelers are indifferent to very small intervals (e.g. 1 minute). Hence, it is safe to assume a uniform departure pattern within the specified time interval without losing much realism.

In the bi-level optimization approach, the upper-level problem is an ordinary least-squares (OLS) problem, which is to estimate the TDOD demand based on given link-flow proportions. The link-flow proportions are in turn generated from the dynamic traffic network loading problem at the lower level, which may be solved by a simulation-based DTA procedure (in this case we use the DYNASMART-P software) *(Zhou, 2004)*. The process is iterated until convergence in the reduction of root mean squared errors (RMSE) of the estimated link-flows is achieved.

The upper-level problem is a weighted multi-objective optimization problem. A mathematical programming platform AMPL is used with the solver KNITRO suited for large-scale non-linear problems *(Waltz and Plantenga, 2009)*. The solver KNITRO utilizes an interior point/conjugate graduate algorithm in order to converge to the optimum solution in a time-efficient manner *(Nocedal and Wright, 2006)*. *The first objective* is to minimize the squared deviations between the simulated flows $M_{l,t}$ and the observed flows $O_{l,t}$ for all observation links $l \in L$ and simulation time intervals $t \in T$.

$$\min_{d_{i,j,h}} (1-w) \left(\sum_{l=1}^{L} \sum_{t=1}^{T} [M_{l,t} - O_{l,t}]^2 \right) + (w) \left(\sum_{i=1}^{I} \sum_{j=1}^{J} \left[\left\{ \sum_{h=1}^{H} d_{i,j,h} \right\} - \delta_{i,j} \right]^2 \right) \quad (3\text{-}8)$$

$$\text{subject to } d_{i,j,h} \geq 0, \quad \forall i,j,h$$

where

L	: The set of observation links,
l	: The index for observation links; $l \in L$,
T	: The set of simulation time intervals,
t	: The index for simulation time intervals; $t \in T$,
h	: The set of departure time intervals,
H	: The index for departure time intervals; $h \in H$,
I	: The set of origins,
i	: The index for origins; $i \in I$,
J	: The set of destinations,
j	: The index for destinations; $j \in J$,
$d_{i,j,h}$	: Time-dependent OD flow from origin $i \in I$ to destination $j \in J$ at the time interval $h \in H$
$\delta_{i,j}$	: The static OD flow from origin $i \in I$ to destination $j \in J$
$p_{i,j,h,l,t}$	: The proportion of demand for origin i, destination j, at departure time h, observed on link l, at simulation/observation time t.

The second objective is to minimize the squared deviations between the sums of the time-dependent OD flows $d_{i,j,h}$ over the departure time intervals $h \in H$ and static OD flows $\delta_{i,j}$ for all OD pairs $i \in I$ and $j \in J$. It must be noted that $d_{i,j,h}$'s are the decision variables of this problem and the outputs of our estimation problem.

The simulated flows $M_{l,t}$ are solved by the lower-level problem and are a function of the decision variables $d_{i,j,h}$ such that $M_{l,t} = \sum_{i,j,h} p_{i,j,h,l,t} d_{i,j,h}$. $p_{i,j,h,l,t}$ is the so-called link proportion, which describes the fraction of OD flow $d_{i,j,h}$ on the link flow $M_{l,t}$. The two stopping criteria used in this methodology are the root mean squared errors for demand and observations *(Alibabai and Mahmassani, 2008)*:

$$RMSE_{Demand} = \sqrt{\frac{\sum_{i=1}^{I}\sum_{j=1}^{J}\left[\left\{\sum_{h=1}^{H}d_{i,j,h}\right\}-\delta_{i,j}\right]^2}{IJH-1}} \qquad (3\text{-}9)$$

$RMSE_{Demand}$ is the measure of error for the deviation between the new time-dependent demand matrix and the original static demand matrix.

$$RMSE_{Flows} = \sqrt{\frac{\sum_{l=1}^{L}\sum_{t=1}^{T}[M_{l,t}-O_{l,t}]^2}{LT-1}} \qquad (3\text{-}10)$$

$RMSE_{Flows}$ is the measure of error for the deviation between the simulated and the observed link flows.

Figure 3-6 illustrates the conceptual relationship between two criteria used in the optimization process. Since the original static OD matrix (left circle in Figure 3-6) typically does not agree well with the actual observations (right circle), our goal is to find a new time-dependent matrix (middle circle) whose resulting traffic flows are well matched with the observed traffic flows, but at the same time not deviating too much from the original static matrix, which was used as a seed for the new matrix. The final new time-dependent OD matrix is therefore obtained by minimizing both $RMSE_{Flows}$ and $RMSE_{Demand}$.

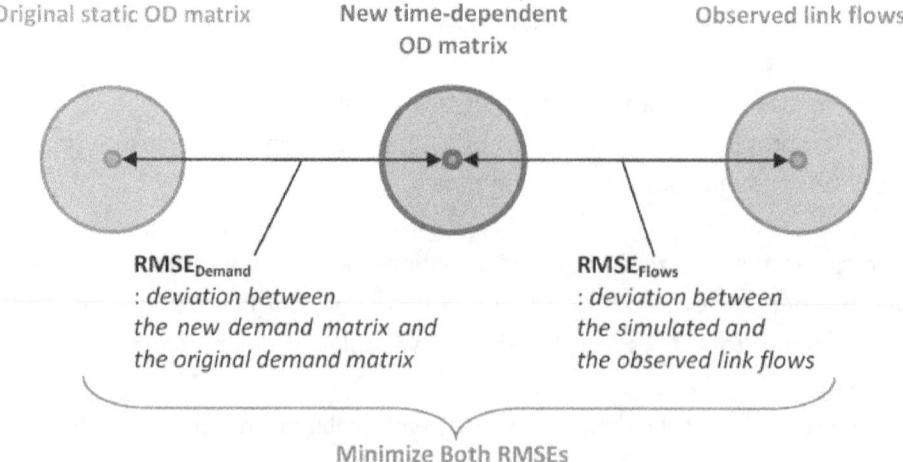

Figure 3-6. Two Criteria in the Optimization Process

3.2.1.2 Estimation Results

This section discusses the estimation results for the base-case time-dependent OD matrix. For each network, we present the convergence pattern of the optimization process for obtaining the final OD matrix and the resulting time-dependent demand profile. For validation purposes, we compare the simulated link flows obtained from the estimated OD matrix to the observed link count data for selected links.

Long Island, NY

Table 3-4 shows the estimation results for the Long Island network. The first two columns represent the number of single and high-occupancy passenger car trips after each iteration. The last two columns show the RMSE values that are discussed in the previous section (i.e., $RMSE_{Demand}$ and $RMSE_{Flows}$).

In the first row, the results associated with the historical OD matrix is also presented for comparison. After the first iteration, $RMSE_{Demand}$ increases from zero because the new time-dependent OD matrix (in the second row), which is created based on the historical OD matrix, is the result of the optimization process that is not only minimizing $RMSE_{Demand}$ but also minimizing $RMSE_{Flows}$. However, after iteration 2, the error does not increase dramatically and always stays below 0.2.

$RMSE_{Flows}$ has decreased 34% after the first iteration and has been continuously decreasing for 8 more iterations. The rate of decrease is decreasing, which implies convergence. This means that the real-world link count observations are matched better with the simulation results produced by the new time-dependent OD matrix than with the historical OD matrix.

As a link-level validation, the simulated and observed link counts are compared for several selected links. Simulated results based on the estimated time-dependent OD matrix are compared with the actual observations, which are collected during the time period (6 am – 11 am) that corresponds to the demand horizon used for the OD matrix estimation. Figure 3-7 displays the cumulative number of vehicle counts (left column) and the 15-minute aggregated vehicle counts (right column) for two selected links, respectively. Overall, link-level comparisons show good agreements.

As a network-wide validation, the overall OD demand pattern is also compared. Figure 3-8 presents the temporal distributions of SOV trips of the historical OD matrix (denoted by "Old SOVs") and the most up-to-date time-dependent OD matrix (denoted by "New SOVs"). Similarly, Figure 3-9 shows the temporal distributions of HOV trips for the historical OD matrix (denoted by "Old HOVs") and the most up-to-date time-dependent OD matrix (denoted by "New HOVs").

Table 3-4. RMSE Values for the Long Island Network

	Number of Trips		RMSE Values	
	SOV*	HOV*	RMSE$_{Demand}$	RMSE$_{Flows}$
Historical OD matrix	1,478,829	608,064	0**	288.775
New time-dependent OD matrix after **Iteration 1**	1,480,138	609,081	0.135	190.654
New time-dependent OD matrix after **Iteration 2**	1,471,142	605,251	0.158	165.129
New time-dependent OD matrix after **Iteration 3**	1,466,169	603,111	0.157	153.852
New time-dependent OD matrix after **Iteration 4**	1,463,592	601,841	0.162	146.108
New time-dependent OD matrix after **Iteration 5**	1,461,663	601,204	0.156	120.654
New time-dependent OD matrix after **Iteration 6**	1,491,665	613,658	0.183	120.428
New time-dependent OD matrix after **Iteration 7**	1,468,132	604,068	0.152	120.030
New time-dependent OD matrix after **Iteration 8**	1,464,196	602,394	0.163	119.996
New time-dependent OD matrix after **Iteration 9**	1,474,278	606,608	0.139	118.849

* *SOV: Single-occupancy vehicle, HOV: High-occupancy vehicle*

** *Deviation is zero because RMSE$_{Demand}$ in this case represents the deviation between the static OD matrix and itself.*

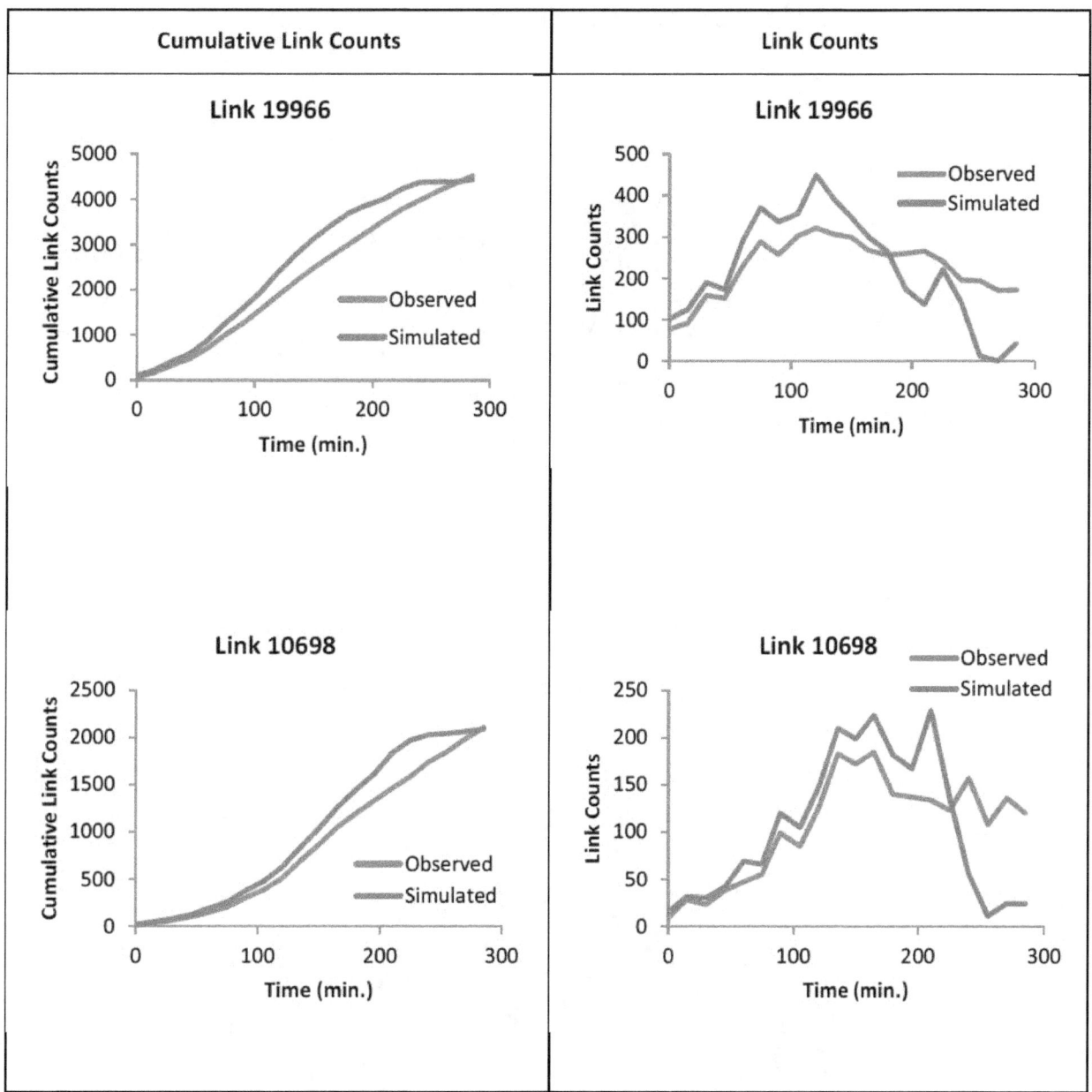

Figure 3-7. Observed and Simulated Counts on Selected Links (Long Island)

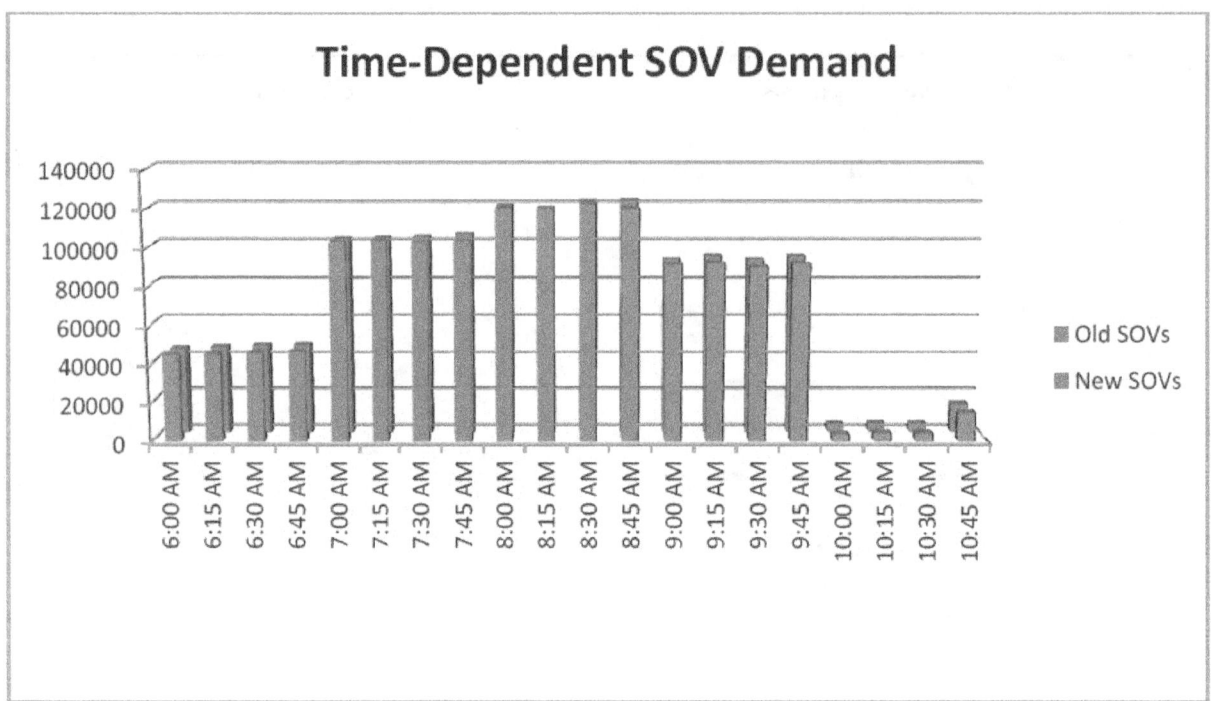

Figure 3-8. Temporal Distribution of SOV trips for the Long Island Network

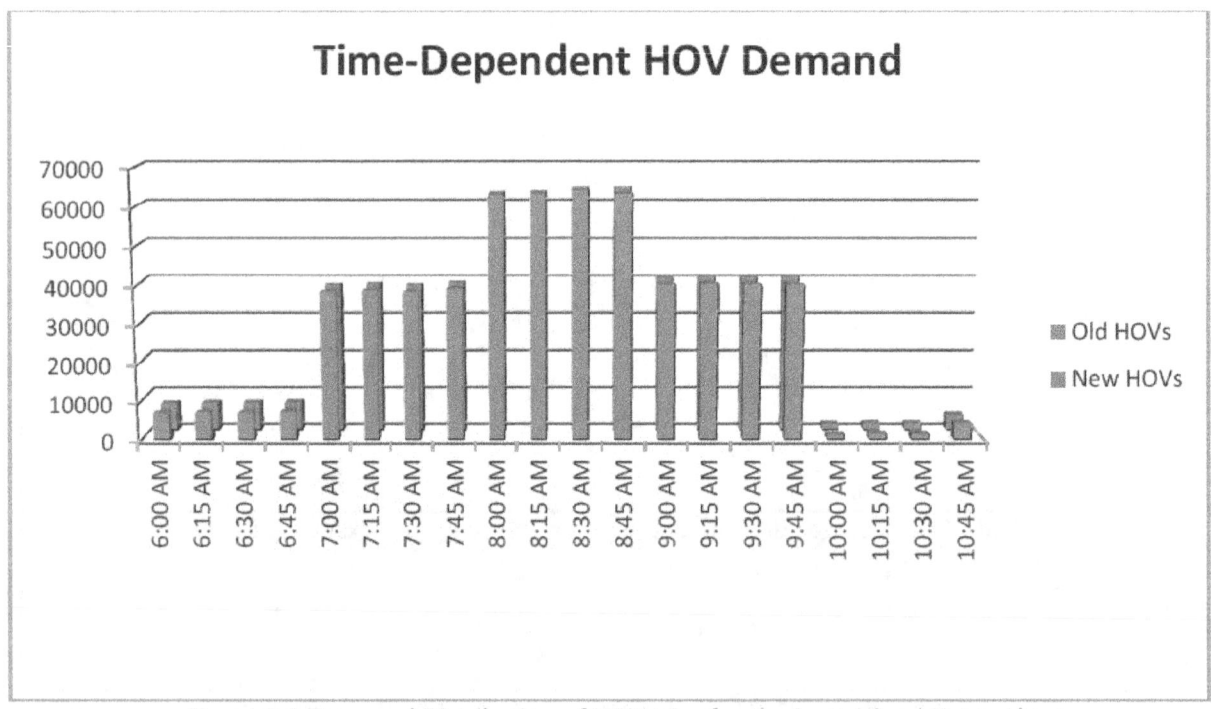

Figure 3-9. Temporal Distribution of HOV trips for the Long Island Network

Chicago, IL

Table 3-5 provides the estimation results for the Chicago network. The results are presented in the same manner as the Long Island network above. The first column shows the number of single-occupancy passenger car trips after every iteration (HOV is not available for the Chicago network). The last two columns show the RMSE values. The $RMSE_{Demand}$ is stabilizing around 0.049 and $RMSE_{Flows}$ decreases after 3 iterations.

Table 3-5. RMSE Values for the Chicago Network

	Number of Trips	RMSE Values	
	SOV^*	$RMSE_{Demand}$	$RMSE_{Flows}$
Historical OD matrix	4,145,413	0^{**}	228.759
New time-dependent OD matrix after **Iteration 1**	4,179,062	0.044	219.148
New time-dependent OD matrix after **Iteration 2**	4,157,199	0.049	217.739
New time-dependent OD matrix after **Iteration 3**	4,141,043	0.049	217.030

* SOV: Single-occupancy vehicle
** Deviation is zero because $RMSE_{Demand}$ in this case represents the deviation between the static OD matrix and itself.

Figure 3-10 displays the cumulative number of vehicle counts (left column) and the 15-minute aggregated vehicle counts (right column) for two selected links between 5 am and 12 am, respectively. Figure 3-11 presents the temporal distributions of SOV trips of the original static OD matrix (denoted by "Old SOVs") and the most up-to-date time-dependent OD matrix (denoted by "New SOVs").

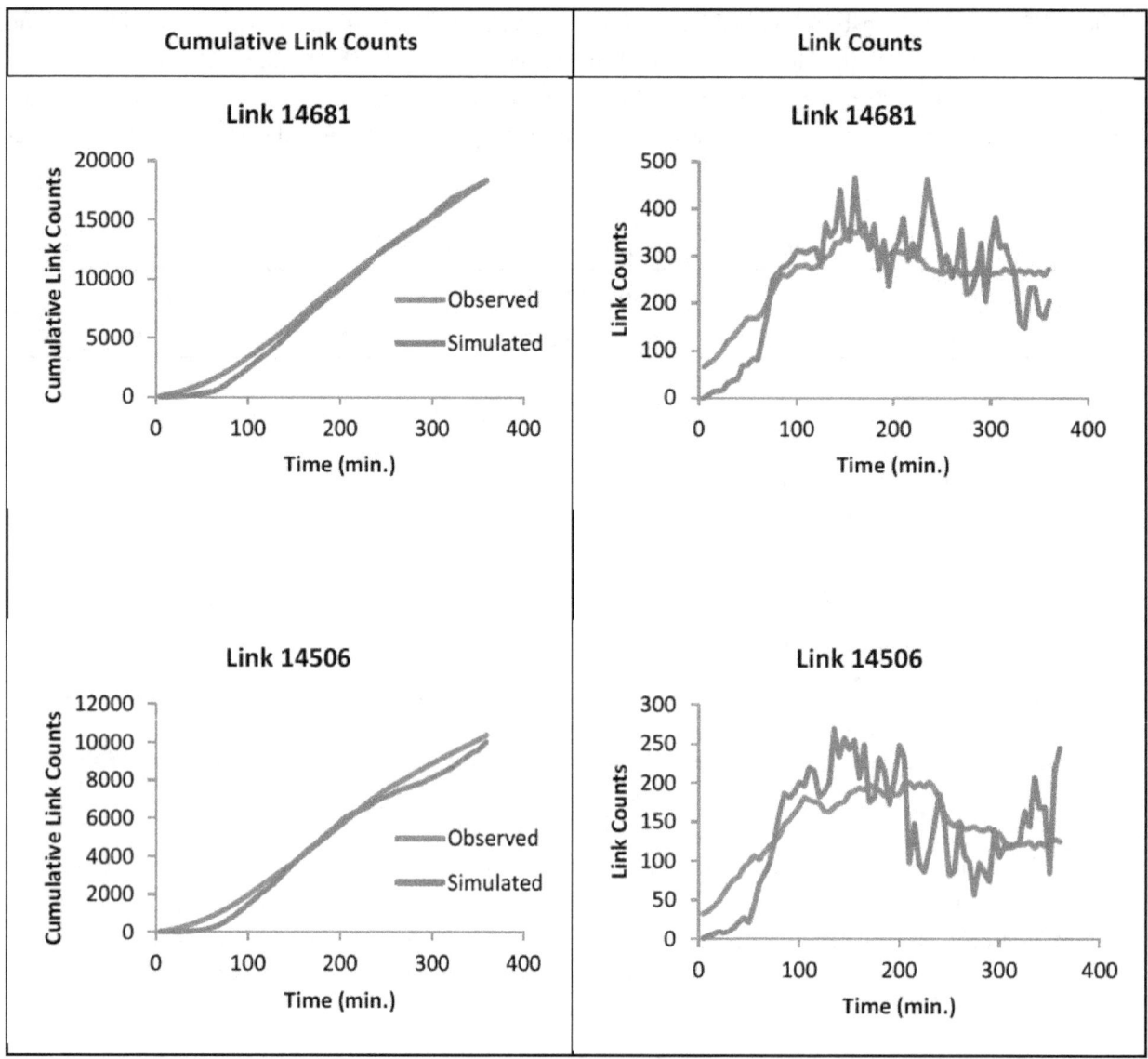

Figure 3-10. Observed and Simulated Counts on Selected Links (Chicago)

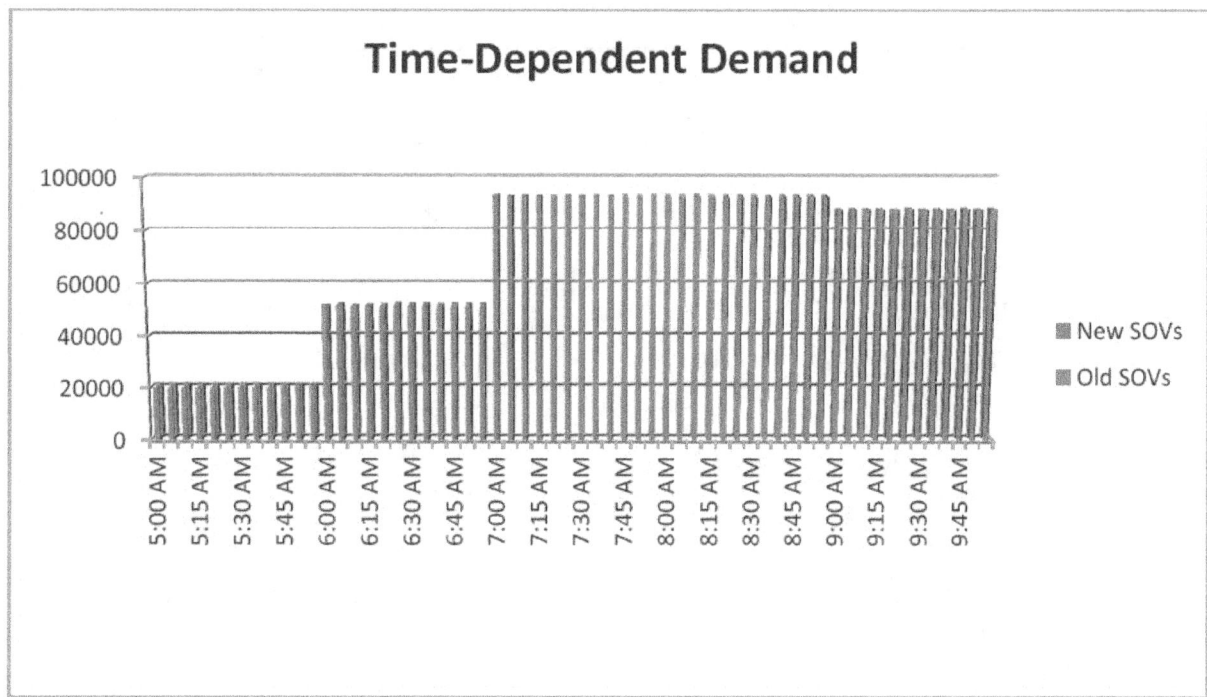

Figure 3-11. Temporal Distribution of trips for the Chicago Network

Salt Lake City, UT

Table 3-6 provides the estimation results for the Salt Lake City network. The static demand matrix was obtained from the Wasatch Front Regional Council, including both SOV and HOV trips. The results are presented in the same manner as the Long Island network above. The first two columns show the number of single-occupancy and high-occupancy passenger car trips after every iteration, respectively. The last two columns show the RMSE values. The RMSE for demand increases at first, but then decreases and stabilizes after a few more iterations. The RMSE for flows decreases continuously.

Figure 3-12 displays the cumulative number of vehicle counts (left column) and the 15-minute aggregated vehicle counts (right column) for a selected link between 6 am and 10 am, respectively. Figure 3-13 and Figure 3-14 present the temporal distributions of SOV and HOV trips of the historical OD matrix (denoted by "Old SOVs" and "Old HOVs" respectively) and the most up-to-date time-dependent OD matrix (denoted by "New SOVs" and "New HOVs" respectively).

Table 3-6. RMSE Values for the Salt Lake City Network

	Number of Trips		RMSE Values	
	SOV*	HOV*	$RMSE_{Demand}$	$RMSE_{Flows}$
Original Static OD matrix	772,017	206,756	0**	296.257
New time-dependent OD matrix after **Iteration 1**	820,145	220,182	0.052	278.861
New time-dependent OD matrix after **Iteration 2**	780,903	206,561	0.062	278.303
New time-dependent OD matrix after **Iteration 3**	746,465	195,893	0.061	277.412
New time-dependent OD matrix after **Iteration 4**	720,380	187,965	0.061	276.547

* SOV: Single-occupancy vehicle, HOV: High-occupancy vehicle

** Deviation is zero because $RMSE_{Demand}$ in this case represents the deviation between the static OD matrix and itself.

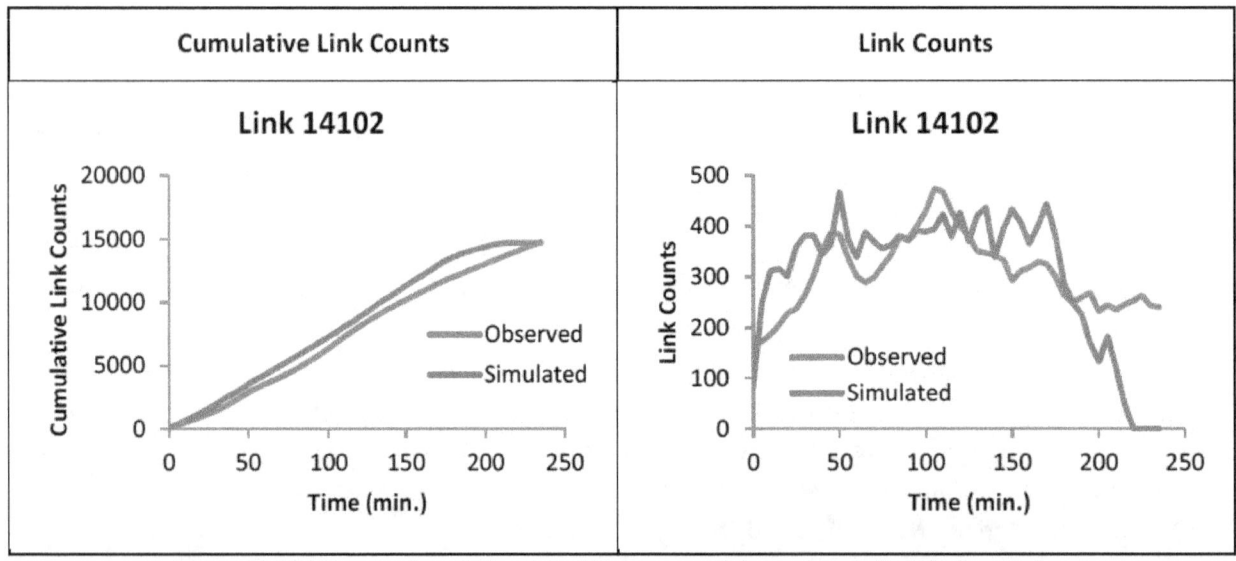

Figure 3-12 Observed and Simulated Counts on Selected Link (Salt Lake City)

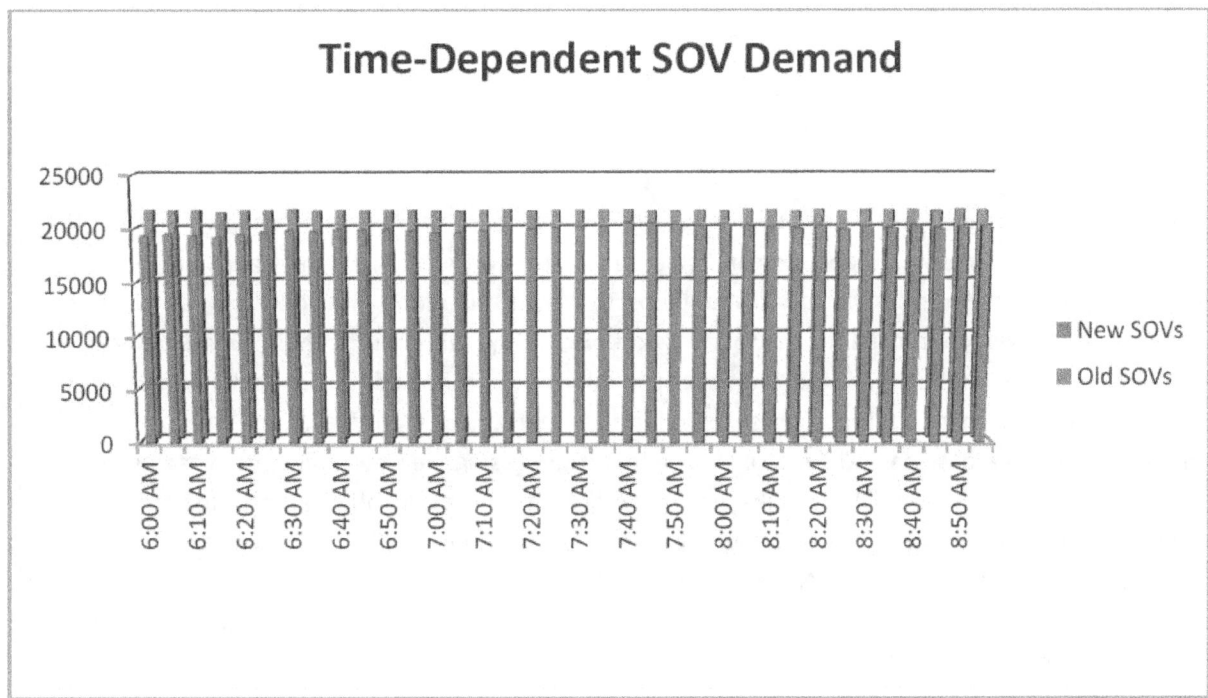

Figure 3-13. Temporal Distribution SOV of trips for the Salt Lake City Network

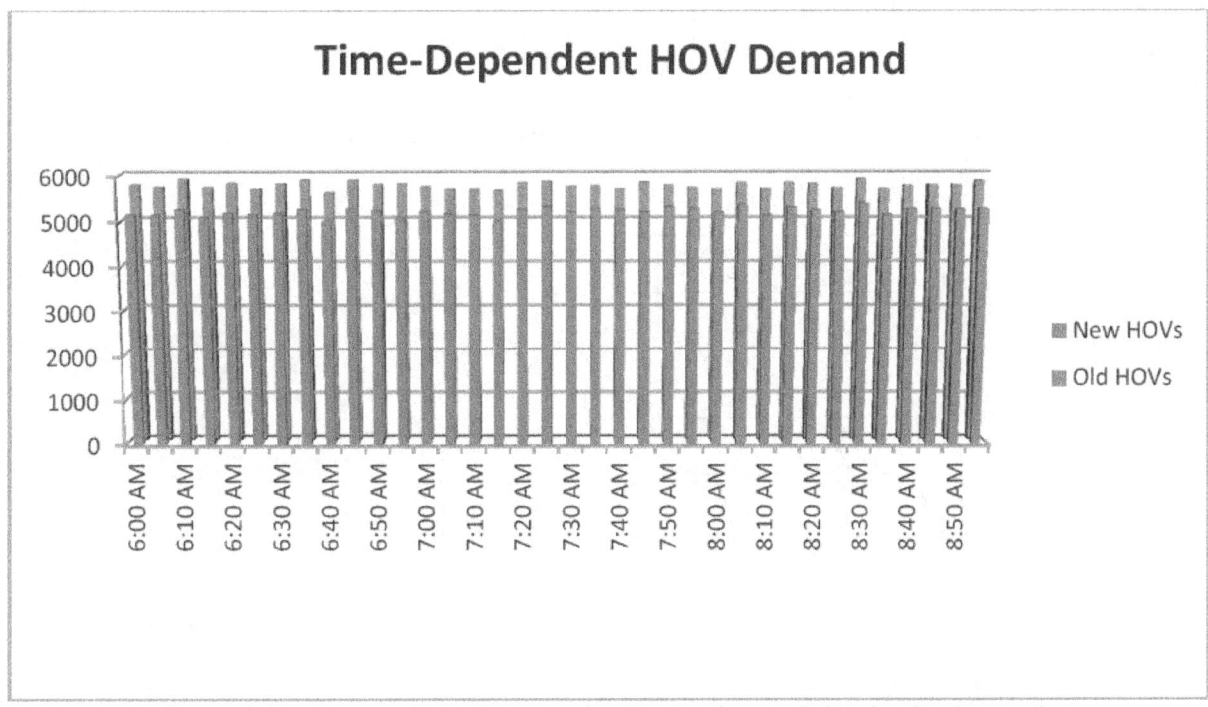

Figure 3-14. Temporal Distribution of HOV trips for the Salt Lake City Network

3.3 Validation of Weather Sensitive DYNASMART-P

3.3.1 Validation Procedure for Weather Specific Simulations

This section discusses the updated results for the validation of the weather-sensitive DYNASMART-P model. After the supply-side and demand-side parameters are obtained, the capability of capturing weather effects on the traffic flows is tested by performing simulations with specific weather scenarios. The Chicago network is chosen for the test network. First, days with rain or snow events between 5AM and 10AM are identified and the traffic observations are collected for each identified day. The corresponding weather conditions are specified in the weather.dat input files. Each weather scenario is simulated with the base-case OD matrix with and without using weather adjustment factors (WAF) in DYNASMART. Then the simulated results are compared with the actual observations under the weather condition the weather.dat is representing. The main focus is to see whether using weather features (i.e., weather.dat and WAF.dat) in DYNASMART produce more realistic traffic flows and speeds, that is, resemble real-traffic conditions under weather better than when running the simulation without considering any weather effect.

Two measures of error have been used: $RMSE_{Flows}$ and $RMSE_{Speeds}$. $RMSE_{Flows}$ represents the discrepancy between the observed and simulated link counts for all time periods for all links. Similarly, $RMSE_{Speeds}$ represents the discrepancy between the observed and simulated link speed for all time periods for all links.

The expectation is that scenario runs without weather adjustment factors would have higher errors than the ones with weather adjustment factors.

$$RMSE_{Flows} = \sqrt{\frac{\sum_{l=1}^{L}\sum_{t=1}^{T}[M_{l,t}-O_{l,t}]^2}{LT-1}} \qquad (3\text{-}11)$$

$$RMSE_{Speeds} = \sqrt{\frac{\sum_{l=1}^{L}\sum_{t=1}^{T}[MS_{l,t}-OS_{l,t}]^2}{LT-1}} \qquad (3\text{-}12)$$

where $M_{l,t}$ is the simulated link flow, whereas $O_{l,t}$ is the observed link flow on link l at time t. Similarly, $MS_{l,t}$ is the simulated speed, and $OS_{l,t}$ is observed link speed on link l at time t.

3.3.2 Validation Results

The following results are based on the analysis for the initial test using a snow scenario observed on January 7, 2010 in Chicago.

Table 3-7 presents the RMSE values for the link flows (counts) and the link speed as discussed above. For the link speed, when the discrepancy between simulation and observation results is much smaller with weather specific parameters (i.e., lower $RMSE_{Speeds}$ for "With Weather Features"). In other words, the use of the weather adjustment factors captures the weather effect on the road traffic thereby producing more realistic simulation results. Similarly, for the link counts the equivalent pattern is observed, that is, the counts are matched better in the simulation by using weather features. The overall experiment results reveal that the weather-sensitive TrEPS indeed has the ability to model the effect of weather conditions. The extent to which the simulation results accurately reflect the actual traffic under weather conditions, however, depends not only on the supply-side adjustment (e.g. WAF) but also on the demand-side adjustment such as the use of OD matrices that are explicitly estimated for specific weather conditions.

Table 3-7. RMSE Values for the selected snow scenario

SNOW Scenario: 2010-01-07 (Chicago)			
RMSE Speeds		RMSE Flows	
With Weather Features	Without Weather Features	With Weather Features	Without Weather Features
22.6939	35.5554	53.2358	67.3506

For a link-level comparison, Figure 3-15 presents observed and simulated speeds with and without weather specific parameters on a selected link. Figure 3-16 presents observed counts vs. simulated counts with and without weather specific parameters on a selected link. In the link-level comparisons, it is observed that simulation results that consider the snow effects are closer to the actual traffic conditions than those that ignore the weather effects.

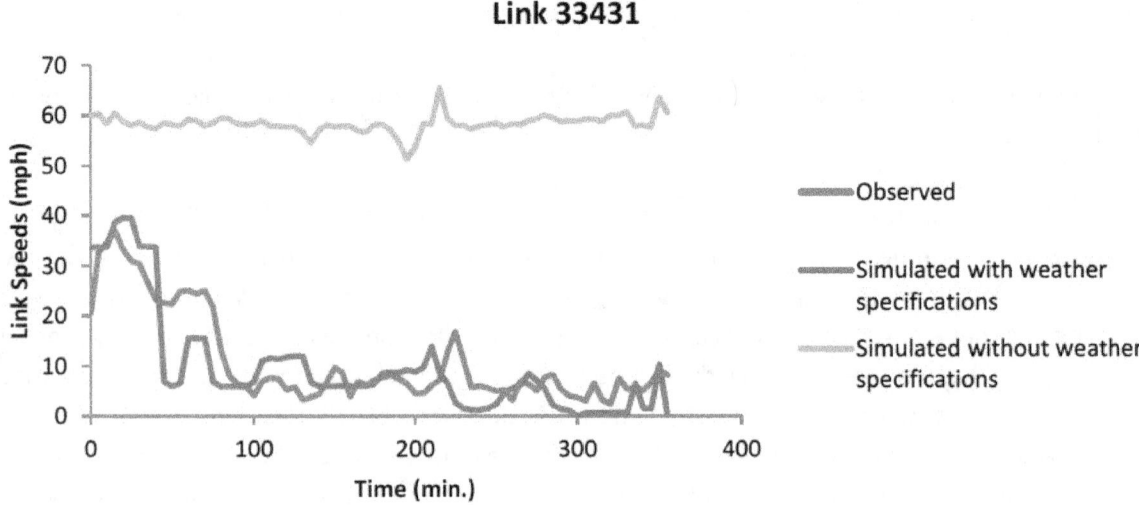

Figure 3-15. Observed and Simulated Speeds on a Selected Link (Chicago)

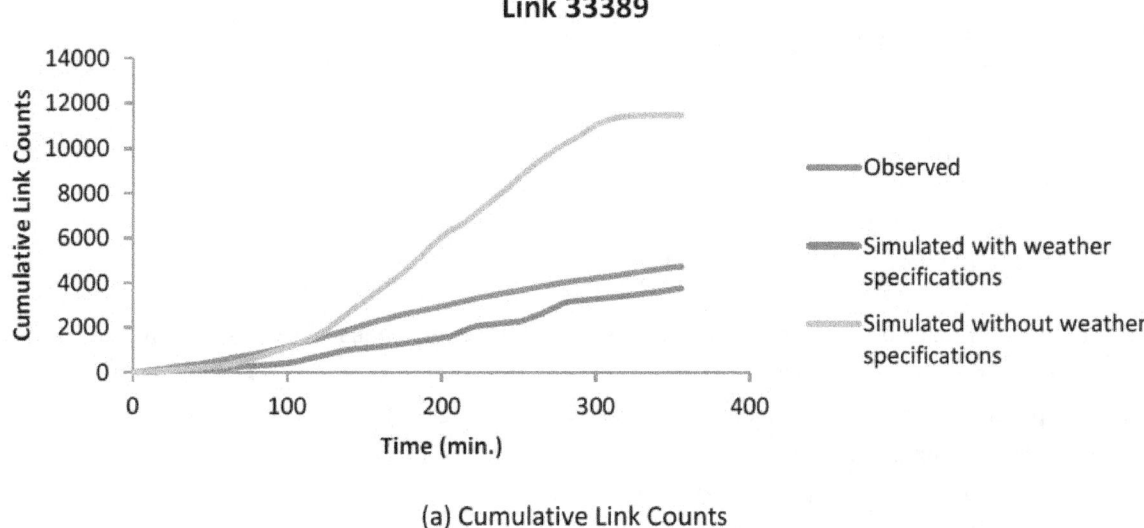

(a) Cumulative Link Counts

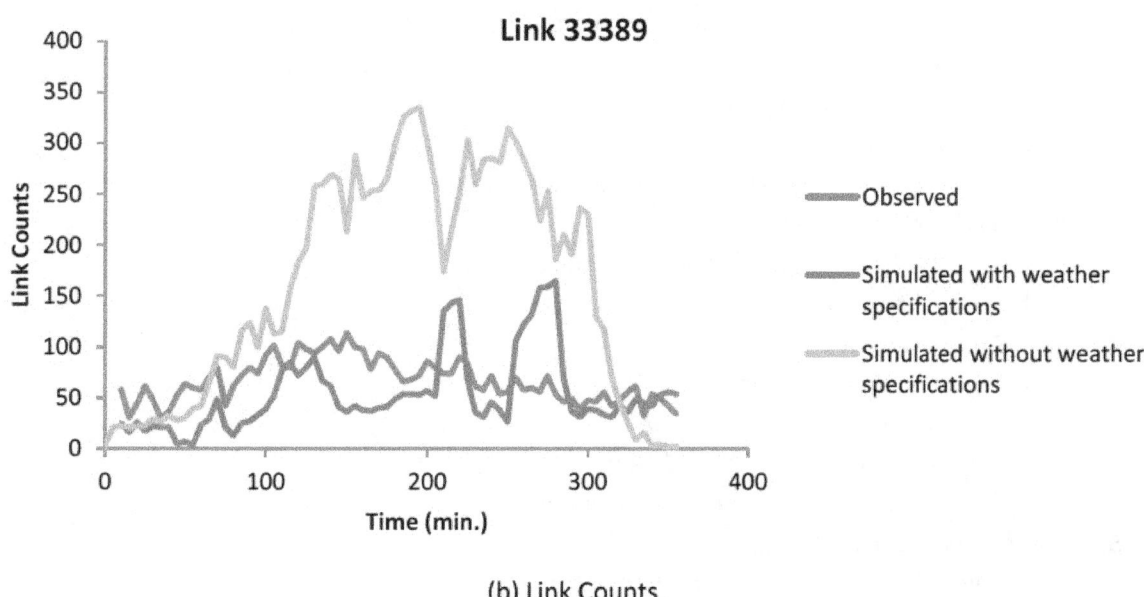

(b) Link Counts

Figure 3-16. Observed and Simulated Counts on a Selected Link (Chicago)

4. Identification of Existing WRTM Strategies

4.1 Background

In 2011, FHWA released a report (FHWA-JPO-11-086) which provides a comprehensive overview of weather-responsive traffic management practices, focusing on existing strategies, their benefits, and how to implement and evaluate them (Gopalakrishna et al, 2011). Following is the categorization of existing WRTM strategies suggested in that report:

1. Motorist Advisory, Alert and Warning Systems
    a. Passive Warning Systems
    b. Active Warning Systems
    c. Pre-Trip Road Condition Information and Forecast Systems
    d. En-Route Weather Alerts and Pavement Condition Information
2. Speed Management Strategies
    a. Speed Advisories
    b. Enforceable Speed Limits/Variable Speed Limits
3. Vehicle Restrictions Strategies
    a. Size/Height/Weight/Profile Restrictions
    b. Tire Chains/Alternate Traction Devices
4. Route Restrictions Strategies
    a. Lane-Use Restrictions
    b. Parking Restrictions
    c. Access Control and Facility Closures
    d. Contraflow/Reversible Lane Operations
5. Traffic Signal Control Strategies
    a. Vehicle Detector Configuration
    b. Vehicle Clearance Intervals
    c. Interval and Phase Duration Settings
    d. Traffic Signal Coordination Plans
    e. Ramp Control Signals/Ramp Metering
6. Traffic Incident Management
    a. Full Function Service Patrols/Courtesy Patrols
    b. Wrecker Response Contracts
    c. Quick Clearance Policies
7. Personnel/Asset Management
8. Agency Coordination

4.2 Identifying Existing WRTM Strategies Used By Selected Agencies

4.2.1 Survey Design

To identify the existing WRTM strategies, a survey has been conducted for agencies in the study sites. Six different categories of WRTM strategies were identified and respondents were asked to indicate which categories of strategies are currently used in response to the inclement weather conditions. Table 2.1 shows these categories and the descriptions of the corresponding strategies.

Table 4-1. Six Categories of WRTM Strategies

No.	WRTM Strategies	Category
1	Display weather information or warning on VMS	Advisory VMS
2	Display road closure information on VMS (e.g. snowplowing operations, flooded area, icy road)	Mandatory VMS
3	Adjust speed limits in response to prevailing weather conditions	Speed Management
4	Modify signal timing plans to improve traffic conditions under inclement weather	Signal Control
5	Modify ramp metering timing plans in response to prevailing weather conditions	Ramp metering
6	Use demand management scheme to reduce the overall volume under adverse weather conditions (e.g. restriction on single occupancy vehicle, restriction on auto-mode, impose higher tolls using certain roads)	Demand Management

4.2.2 Survey Results

Based on the survey results, we obtained the information on the existing WRTM strategies implemented by different agencies. Table 2.2 provides a summary of the responses from agencies who revealed the use of WRTM strategies in their traffic operations. The following is a brief description of the survey results and comments from agencies:

- The Chicago DOT implements VMS-related strategies in coordination with the Illinois DOT using both advisory and mandatory VMS to distribute warning/restriction information to drivers. Depending on severity roadways, segments may be closed, or transit services may be re-routed in coordination with CTA.

- The Utah DOT implements advisory/mandatory VMS, speed management, and signal control strategies (the speed management strategies are only used in planning phase). The agency also produces forecast of future road conditions during weather events that are posted to the CommuterLink website (http://commuterlink.utah.gov) and customized weather forecasts for each maintenance shed during the winter to facilitate planning of snow plow operations (timing, staffing, materials, etc.).
- The Salt Lake City Transportation Division implements WRTM strategies in most of the categories (except the demand management). Those include displaying the weather information and the road closure information on VMS, adjusting speed limit, and modifying signal and ramp metering timing plans.
- The NY State DOT implements a WRTM strategy using VMS to alert the public of road closure due to weather-related situations (e.g. snowplowing operations, flooded area, icy road). Depending on the severity of weather events, NYS Thruway may restrict tandem trailers.
- The NY City DOT implements advisory/road closure warnings over VMS and Highway Advisory Radio (HAR). The agency also implements advisory speed limit on bridges for high winds; signal timing adjustment strategy to manage expected directional demand; and emergency restriction advisory on travel.
- The City of Irvine reveals that no specific traffic management strategies are implemented in adverse weather conditions.

Table 4-2. Survey Response from Agencies

WRTM Strategies	Agencies				
	Chicago	Salt Lake City		New York (Long Island)	
	Chicago DOT	Utah DOT	Salt Lake City Trans. Division	NY State DOT	NY City DOT
Display weather information or warning on VMS	Yes	Yes	Yes	No	Yes
Display road closure information on VMS (e.g. snowplowing operations, flooded area, icy road)	Yes also coordinate with IDOT	Yes	Yes	Yes	Yes
Adjust speed limits in response to prevailing weather conditions	No	Yes but only in planning phase	Yes	No	Yes on bridges for high winds
Modify signal timing plans to improve traffic conditions under inclement weather	No	Yes	Yes	No	Yes to manage expected directional demand
Modify ramp metering timing plans in response to prevailing weather conditions	No	No	Yes	No	No
Use demand management scheme to reduce the overall volume under adverse weather conditions (e.g. restriction on single occupancy vehicle, restriction on auto-mode, impose higher tolls using certain roads)	No	No	No	No NYS Thruway may restrict tandem trailers	Yes

4.3 Recommendation of Candidate WRTM Strategies for This Study

4.3.1 Basis for Recommendation

In this section, we discuss recommended WRTM strategies for the implementation and evaluation using TrEPS models. Based on the survey results identified above, WRTM strategies are selected for two different groups:

- Existing WRTM strategies, i.e., strategies that are currently implemented based on standard operating procedures (e.g., under what conditions which strategy is deployed); and
- New (potential) WRTM strategies, i.e., strategies that have not been adopted yet but could be tested for future extension.

The extent to which TrEPS supports the decision-making procedure in implementing and evaluating WRTM strategies is different between the two groups. The following paragraphs describe these different aspects, which serve as the basis for the recommendation.

For the existing strategies, the TrEPS assists TMC operators in determining when to activate/deactivate them on top of the standard operating procedures or guidelines. By continually simulating and monitoring the predicted traffic conditions, the operators can implement selected WRTM strategies in a proactive manner, i.e., activate the strategies in advance of deteriorating roadway and traffic conditions, and deactivate them when the intervention is no longer needed. In addition to the decision of when to deploy WRTM strategies, decisions of which scenario would provide the most desirable benefits (among different signal timing plans, levels of speed limit adjustment, combinations of different strategies, etc.) can also be supported by the TrEPS. The system prepares a set of simulation input files for the WRTM strategies that meet the operating policies for the prevailing weather condition. The operators then can easily conduct the simulation for each scenario and assess its predicted effectiveness.

For new strategies, the TrEPS will provide the traffic managers with opportunities to test recommended WRTM strategies for the future implementation. As revealed in the recent study (Gopalakrishna et al, 2011), agencies are hesitant to deploy new strategies for fear of liability. Lack of information on the impacts of those strategies on driver behavior, performance and safety hinders the new strategy adoption. Agencies might desire more pilot studies with rigorous evaluations that clearly demonstrate the benefits of different types of WRTM strategies. The TrEPS is especially helpful in this regard. Various new strategies could be simulated during weather events in parallel with actual operations using existing strategies. The predicted traffic conditions under the new strategies would provide the operators with an insight into the potential effects and relative benefits compared to the existing ones. The real-time TrEPS would allow more timely "what-if" analyses than the off-line tools as its prediction is performed based on the prevailing traffic and weather conditions.

4.3.2 Recommended WRTM Strategies for Each Network

As Irvine does not implement WRTM strategies in traffic operations, the recommendations pertain to the other three networks: Chicago, Salt Lake City and New York's Long Island. This section presents an initial recommendation of candidate WRTM strategies for each network (rather than each agency). Tables 2.3 to 2.5 list candidate WRTM strategies for Chicago, Salt Lake City and New York, respectively. Strategies are divided into above-mentioned two groups: existing strategies and new (potential) strategies. The former are identified from the survey results shown in Table 2, and the latter are suggested based on the general guidelines discussed in (Gopalakrishna et al, 2011) and current information on available network facilities.

For implementing and evaluating the existing WRTM strategies, the Chicago network includes the advisory VMS, which may display congestion or risk warning, optional detour information, and suggested speed; and mandatory VMS, which forces vehicles to change their routes because of road closure or other weather-related operations. For the Salt Lake City and New York networks, speed management (variable speed limit) and signal control are included in addition to the advisory and mandatory VMS.

For the purpose of testing new WRTM strategies, we specify variable speed limit, ramp metering for Chicago; ramp metering for Salt Lake City; and demand management for New York as our initial recommendation.

Table 4-3. WRTM Strategy Recommendation for the Chicago network

Recommended WRTM Strategies		Applicable Weather Events/Situations
Existing WRTM Strategies		
Advisory VMS	***Weather Information or Warning VMS***: Display inclement weather conditions and travel risks due to weather so that drivers reevaluate their current routes. ***Optional Detour VMS***: Advise drivers of weather-impacted areas that lie ahead and suggest possible detour paths. ***Speed Reduction Advisory VMS***: Issue speed advisories in response to deteriorating weather conditions.	Heavy rainfall/snowfall, Snow and ice accumulations, limited visibility, flooded area, snow plowing operations, etc.
Mandatory VMS	***Mandatory Detour VMS***: Display road closure information and mandate all vehicles to follow recommended detour paths in the vicinity.	Flooded area, snow plowing operations, etc.
New WRTM Strategies		
Speed Management	***Variable Speed Limit (VSL)***: Adjust speed limits based on prevailing weather conditions and require drivers to slow down accordingly.	Heavy rainfall/snowfall, Snow and ice accumulations, limited visibility, etc.
Ramp Metering	***Ramp Metering***: Adjust ramp metering timing plans to account for lost freeway capacity, slow travel speeds, and increased start-up time at ramp control signals under adverse weather conditions.	Heavy rainfall/snowfall, Snow and ice accumulations, limited visibility, etc.

Table 4-4. WRTM Strategy Recommendation for the Salt Lake City network

Recommended WRTM Strategies		Applicable Weather Events/Situations
Existing WRTM Strategies		
• Advisory VMS	**Weather Information or Warning VMS** : Display inclement weather conditions and travel risks due to weather so that drivers reevaluate their current routes. **Optional Detour VMS** : Advise drivers of weather-impacted areas that lie ahead and suggest possible detour paths. **Speed Reduction Advisory VMS** : Issue speed advisories in response to deteriorating weather conditions.	Heavy rainfall/snowfall, Snow and ice accumulations, limited visibility, flooded area, snow plowing operations, etc.
• Mandatory VMS	**Mandatory Detour VMS** : Display road closure information and mandate all vehicles to follow recommended detour paths in the vicinity.	Flooded area, snow plowing operations, etc.
• Speed Management	**Variable Speed Limit (VSL)** : Adjust speed limits based on prevailing weather conditions and require drivers to slow down accordingly.	Heavy rainfall/snowfall, Snow and ice accumulations, limited visibility, etc.
• Signal Control	**Signal Interval and Phase Duration settings** : Alter the duration and/or sequencing (e.g., min/max green time, amber time and cycle length) of traffic signal phases during inclement weather conditions to account for increases in start-up lost time, reduced travel speeds, and reduced pavement traction.	Heavy rainfall/snowfall, Snow and ice accumulations, limited visibility, etc.
New WRTM Strategies		
• Ramp Metering	**Ramp Metering** : Adjust ramp metering timing plans to account for lost freeway capacity, slow travel speeds, and increased start-up time at ramp control signals under adverse weather conditions.	Heavy rainfall/snowfall, Snow and ice accumulations, limited visibility, etc.

Table 4-5. WRTM Strategy Recommendation for the New York network

Recommended WRTM Strategies		Applicable Weather Events/Situations
Existing WRTM Strategies		
• Advisory VMS	***Weather Information or Warning VMS***: Display inclement weather conditions and travel risks due to weather so that drivers reevaluate their current routes. ***Optional Detour VMS***: Advise drivers of weather-impacted areas that lie ahead and suggest possible detour paths. ***Speed Reduction Advisory VMS***: Issue speed advisories in response to deteriorating weather conditions.	Heavy rainfall/snowfall, Snow and ice accumulations, limited visibility, flooded area, snow plowing operations, etc.
• Mandatory VMS	***Mandatory Detour VMS***: Display road closure information and mandate all vehicles to follow recommended detour paths in the vicinity.	Flooded area, snow plowing operations, etc.
• Speed Management	***Variable Speed Limit (VSL)***: Adjust speed limits based on prevailing weather conditions and require drivers to slow down accordingly.	Heavy rainfall/snowfall, Snow and ice accumulations, limited visibility, etc.
• Signal Control	***Signal Interval and Phase Duration settings***: Alter the duration and/or sequencing (e.g., min/max green time, amber time and cycle length) of traffic signal phases during inclement weather conditions to account for increases in start-up lost time, reduced travel speeds, and reduced pavement traction.	Heavy rainfall/snowfall, Snow and ice accumulations, limited visibility, etc.
New WRTM Strategies		
• Demand Management	***Vehicle Size/Height/Weight/Profile Restrictions***: Restrict specific types of vehicles from using the roadways during specific weather conditions. Vehicles may be restricted by size, height, weight, or profile based on weather conditions. (e.g., tandem trailers) ***Vehicle Mode/Occupancy Restrictions***: Apply restrictions on specific vehicle mode or occupancy (e.g., auto-mode, single occupancy vehicle) to manage high-demand situations safely and effectively during weather events.	High wind, heavy rainfall/snowfall, etc.

5. Evaluation Approaches to Assess Benefits of WRTM

5.1 Purpose

This chapter provides an overview of various performance measures that can be used to evaluate the effectiveness and benefits of WRTM strategies. The methodologies and procedures would differ between evaluating simulated strategies and evaluating actually deployed strategies. As depicted in Figure 5-1, WRTM strategies that are simulated using TrEPS models are evaluated based on the simulation output such as vehicle trajectory data. The information on the predicted benefits is then used to determine whether the given strategies can be actually deployed (for on-line TrEPS implementation); or to compare various scenarios for developing new WRTM strategies for future use (for off-line simulation experiments). On the other hand, WRTM strategies that are deployed in the real-world road network are evaluated based on actual traffic observations such as detector data. The purpose of this evaluation is to obtain feedback on the implemented strategies in order to improve and update the WRTM selection and deployment policy. A recent FHWA project *(Gopalakrishna et. al, 2011)* defines a few of the designs for the evaluation of actually deployed strategies that may be appropriate as follows:

Before – After method	Compare data under baseline conditions before deployment and use of the WRTM strategy with post-deployment conditions for same locations under similar weather events.
With–Without method	Compare experimental and control sites that are similar except that the experimental site uses the WRTM strategy and the control site does not.

Although these traditional field-based methods might be more definitive than a simulation-based evaluation approach, these not only require considerable time and effort but also could create liability issues for the implementing agency as well as budgetary issues for deployment and analysis. Also, the accuracy and reliability of the assessment results highly depend on the availability of data including both the traffic data in the desirable format or level of detail and coverage and the weather observations for the weather event of interest.

Under such circumstances, a desirable approach is to use cost-effective simulations to the extent possible, so as to minimize the number of required pilot field test instances needed to evaluate the effectiveness of the WRTM strategies of interest. Thoroughly calibrated and validated TrEPS models can produce the simulation outputs that reflect the real-world traffic conditions and the user responses to the simulated WRTM strategies.

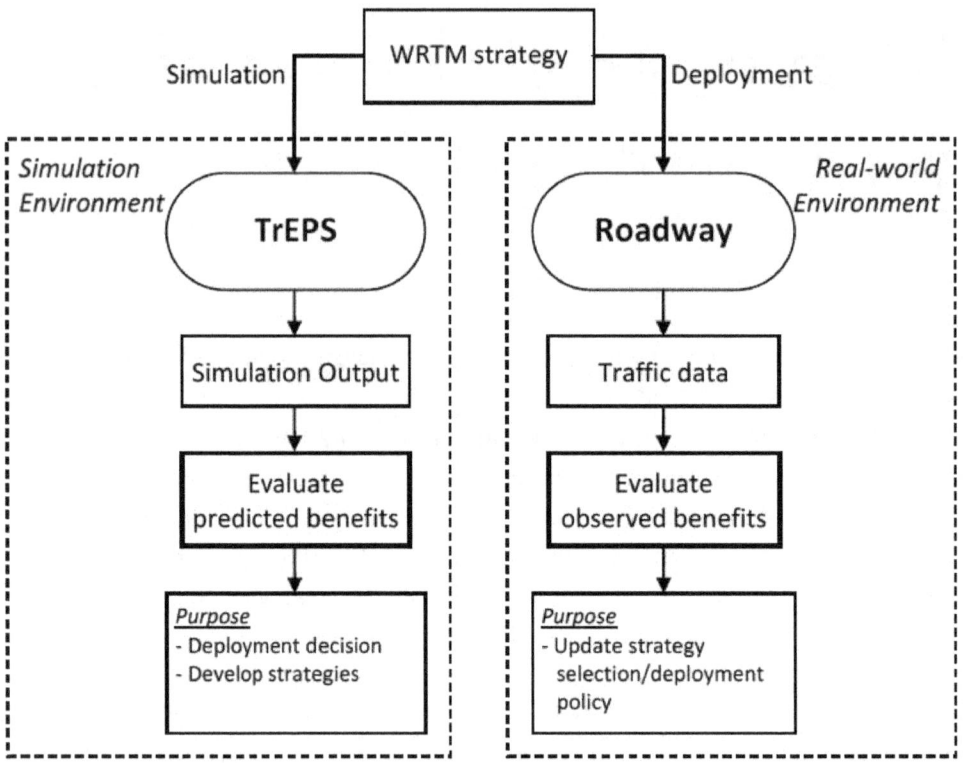

Figure 5-1. Evaluating WRTM strategies in Simulation and Real-world Environments

Figure 5-2 illustrates how TMC operators can evaluate the predicted benefits of a selected WRTM strategy using DYNASMART-X. Based on the current traffic state estimated using the real-time traffic data (RTDYNA), DYNASMART-X predicts two different future conditions; (i) one without any strategy (PDYNA_0) and (ii) the other with an intervention scenario (PDYNA-1). If a user clicks any link on the network, he/she can examine the link performance (e.g., density in this example) from the past to the current time point (blue line in the bottom figure) as well as the future performance without intervention (yellow line) and that with intervention (purple line). The gap between yellow and purple lines shows the potential benefit of applying the given intervention strategy.

Ch.5　　　　　　　　　　　　　　　　Evaluation Approaches to Assess Benefits of WRTM

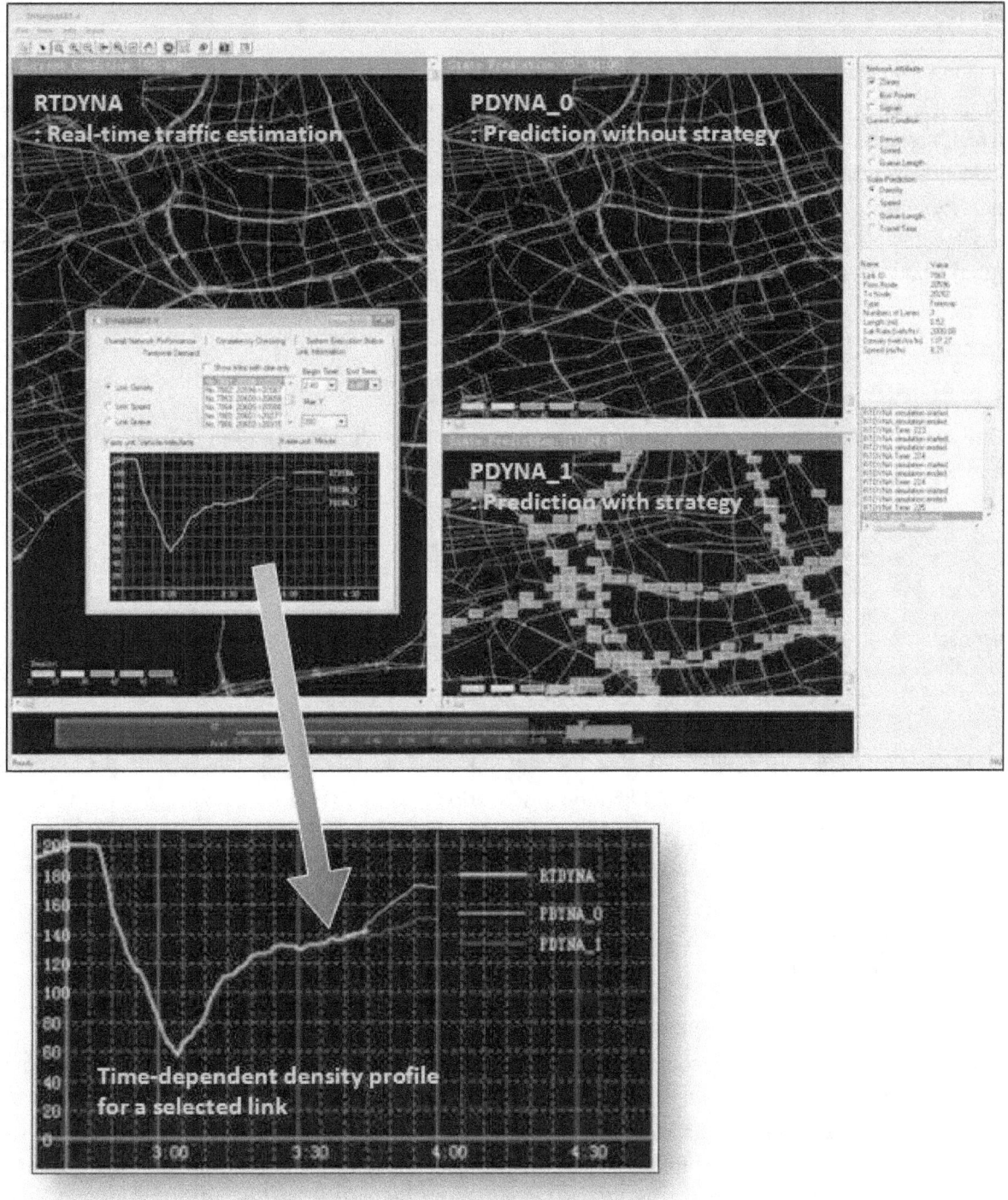

Figure 5-2. DYNASMART-X GUI; Prediction Results with and without Intervention

5.2 Performance Measures

An important advantage of using the TrEPS models for the evaluation of WRTM strategies is the capability of producing a wide variety of performance measures. Based on the vehicle trajectory data, which are the main simulation outputs of the TrEPS model and contain all the detailed movements of individual vehicles, one can extract any mobility-related performance measures (e.g., travel time, speed, density and flow)and analyze them at various levels of detail (e.g., network-level, OD/path-level and link-level). This section provides an overview of various performance measures that are produced from the DYNASMART simulation model.

5.2.1 Network-level Measures

5.2.1.1 Network Throughput

One way to quantitatively measure an overall network throughput over time is using the following metric:

$$\%Accumulated\_Out\_Veh_i^t = \frac{Out\_Veh_i^t}{Tot\_Veh_i^t} \times 100 \qquad (5\text{-}1)$$

Where

$Out\_Veh_i^t$ Accumulated number of vehicles arriving their destinations from time 0 till time t in scenario i

$Tot\_Veh_i^t$ Accumulated total number of vehicles generated and loaded onto the network from time 0 till time t in scenario i

This measure represents how many vehicles have exited the network with respect to the total number of generated vehicles up to every time point and is readily obtained after finishing the simulation. This overall network throughput measure can be used to assess a network-wide WRTM strategy such as demand management.

5.2.1.2 Network Congestion Level(Spatial Distribution)

In order to identify where the congestion forms on the network, it is useful to display traffic attributes on the map view to obtain snapshots at a particular point in time. The DYNASMART graphical user interface (GUI) displays traffic flow parameters (i.e., speed, density and flow) and queue length on each link over the network as shown in Figure 5-3. By sliding a simulation clock time cursor in the GUI, users can examine the MOEs pertaining to different simulation time points. Figure 5-3 (a) shows link speed, where the colors represent different levels of traffic congestion (e.g., red = low speed = congested; green = high speed = uncongested), and Figure 5-3 (b) shows queue lengths.

(a) *Speed* (b) *Queue Length*

Figure 5-3. Visual Representation of Performance Measures for Chicago Network

5.2.1.3 Network Congestion Level (Temporal Distribution)

In addition to examining the state of the system from the map view, it is possible to present the dynamic network traffic state over time on a chart. Figure 5-4 and Figure 5-5 show time-dependent percentages of lane-miles for different density and speed levels, respectively. These charts represent the percentage of lane-miles congested at each time point, where the lane-miles are defined as the sum of the miles for all lanes in each link in the network. This measure is useful for grasping the overall temporal trend of the network congestion level as well as spotting a time point of the onset of congestion. As such, comparing different scenarios (e.g., scenarios with and without intervention) using this measure would provide TMC operators with a useful way of analyzing and quantifying different network congestion levels under various scenarios. As an example, two illustrative scenarios are presented in Figure 5-4 and Figure 5-5. Let Scenario 1 (top) represent a scenario with inclement weather and a certain WRTM strategy to deal with the weather impact, and let Scenario 2 represent a scenario with inclement weather without any intervention strategy. Figure 5-4 and Figure 5-5 show how the tested WRTM strategy improves the network density and speed levels, respectively, as the fraction of line-miles that are heavily congested (e.g., red area) decreases in Scenario 1 compared to Scenario 2.

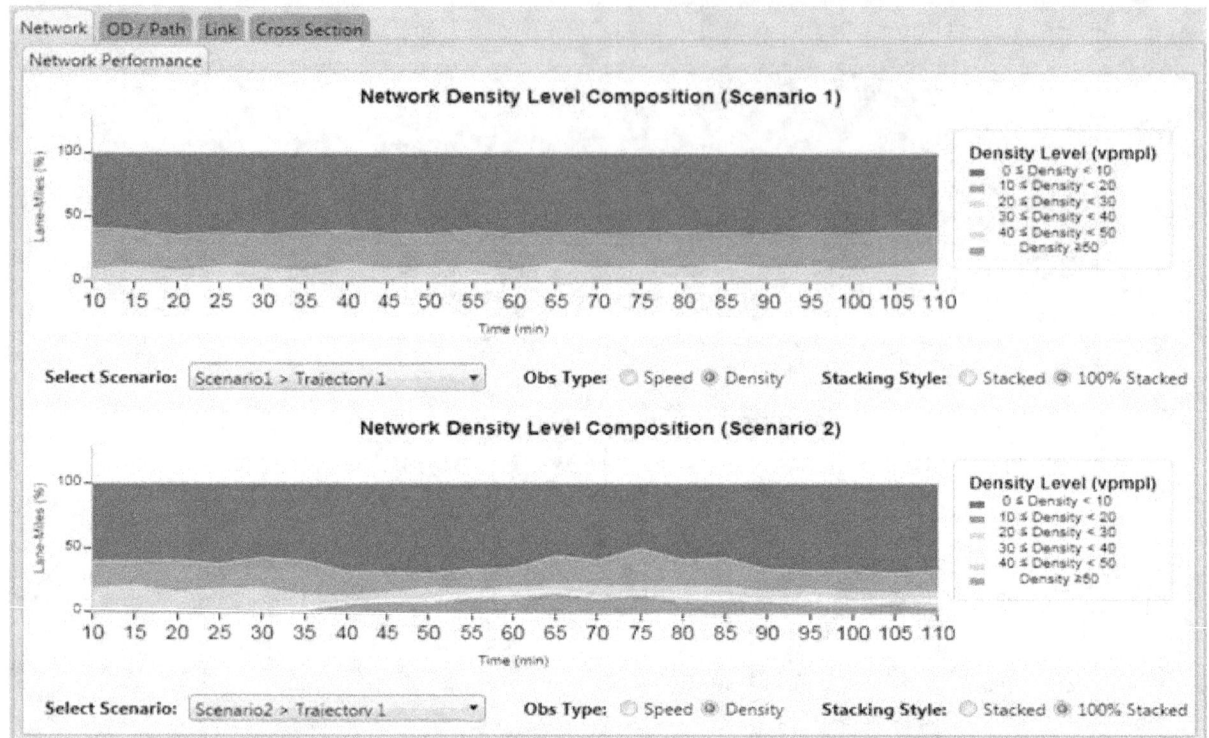

Figure 5-4. Percentage of Lane-miles for Each Density Level (Scenario 1 vs. Scenario 2)

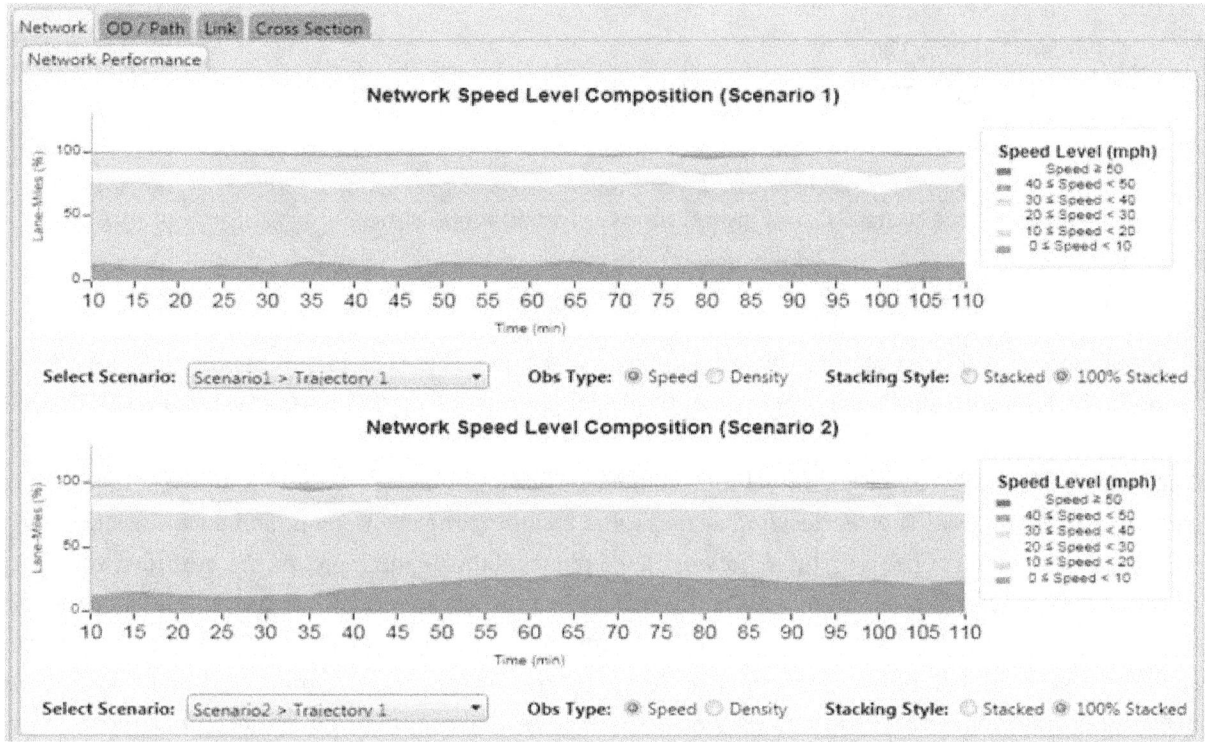

Figure 5-5. Percentage of Lane-miles for Each Speed Level (Scenario 1 vs. Scenario 2)

5.2.2 OD/Path-level Measures

5.2.2.1 Travel Time Distribution (Reliability Measures)

Once the vehicle trajectory data are produced from the TrEPS model as simulation output, the travel time distribution for a certain origin and destination (OD) pair or a certain path can be extracted. One can then estimate various descriptive statistics and reliability measures from the distribution. Figure 5-6 shows a GUI for plotting the travel time distribution for a selected OD pair and presenting the associated statistics (i.e., mean, median, standard deviation and $25^{th}/75^{th}/95^{th}$ percentiles) and reliability measures. The reliability measures include Buffer Index, Planning Time Index, Percent On Time and Misery Index, whose definitions are described as follows:

Buffer Index (%)

$$= \frac{(95^{th} \text{ percentile travel time} - \text{Mean travel time})}{\text{Mean travel time}} \times 100 \qquad (5\text{-}2)$$

Planning Time Index

$$= \frac{95^{th} \text{ percentile travel time}}{\text{Free-flow travel time}} \qquad (5\text{-}3)$$

Percent On Time (%)
= Percentage (Travel time < 1.1 × Median travel time) (5-4)

Misery Index
$$= \frac{(\text{Avg. of the worst 20\% of travel times} - \text{Mean travel time})}{\text{Mean travel time}}$$ (5-5)

A more comprehensive list of reliability performance measures as well as recommended metrics can be found in the literature, particularly the report from SHRP2 Project L03 *(Cambridge Systematics, 2010)*.

One way of effectively comparing travel time characteristics of different scenarios is to use a radar chart as shown in Figure 5-7. The six descriptive measures including mean, median, standard deviation, the 95th percentile, mean of the worst (i.e., longest) 20% of travel times and mean of the best (i.e., shortest) 20% of travel times are presented on axes. By scaling all the axes identically, not only relative ordering within each attribute but also meaningful comparison across attributes can be performed. In general, smaller area indicates better performance (e.g., shorter mean travel time and smaller standard deviation). The radar chart may not be well suited for reliability measures as different metrics have different scales and performance improvement (e.g., distance between values of two scenarios on the same axis) for one metric cannot be directly compared that for another metric.

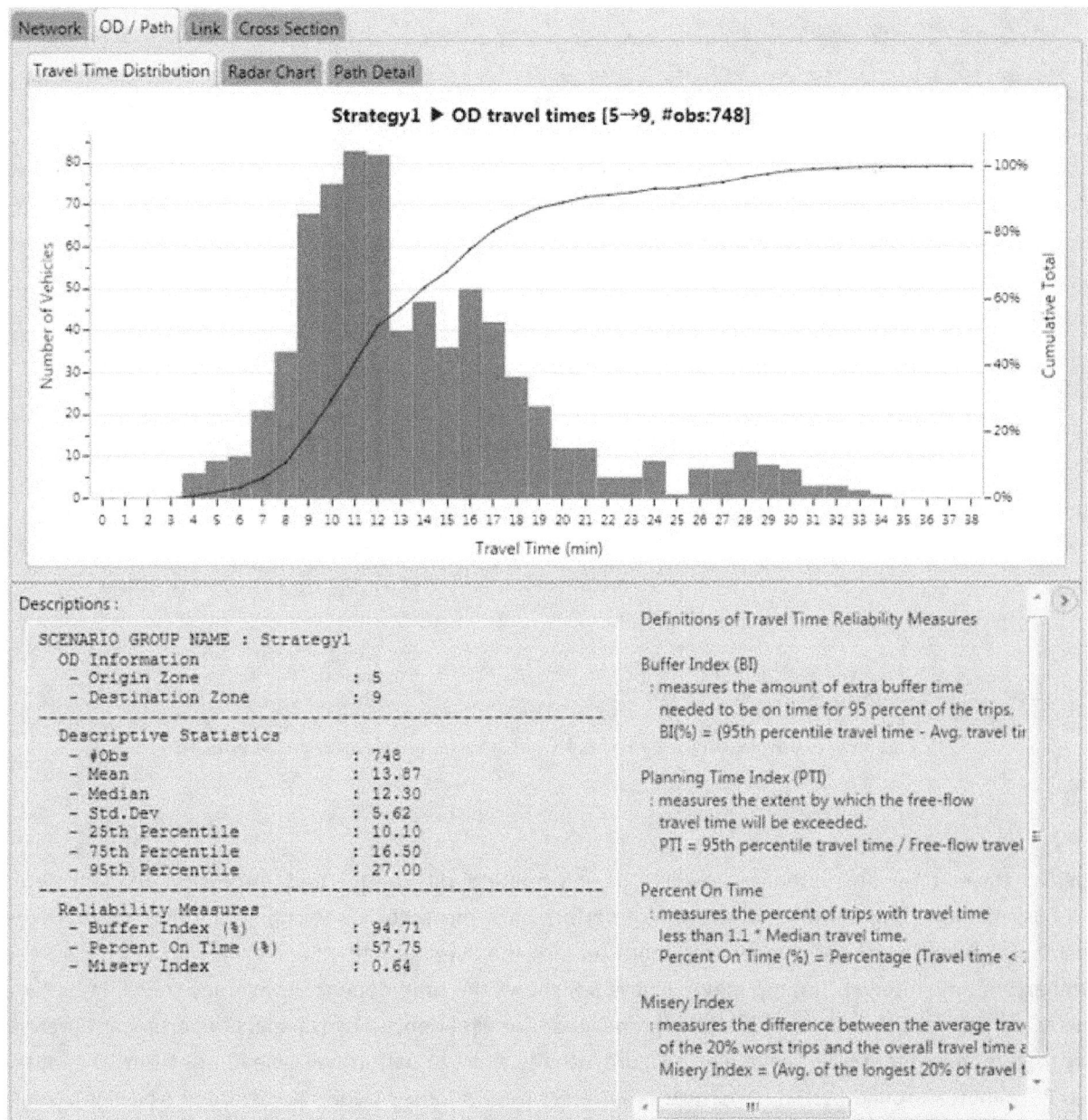

Figure 5-6. OD/Path Travel Time Distribution and Associated Measures

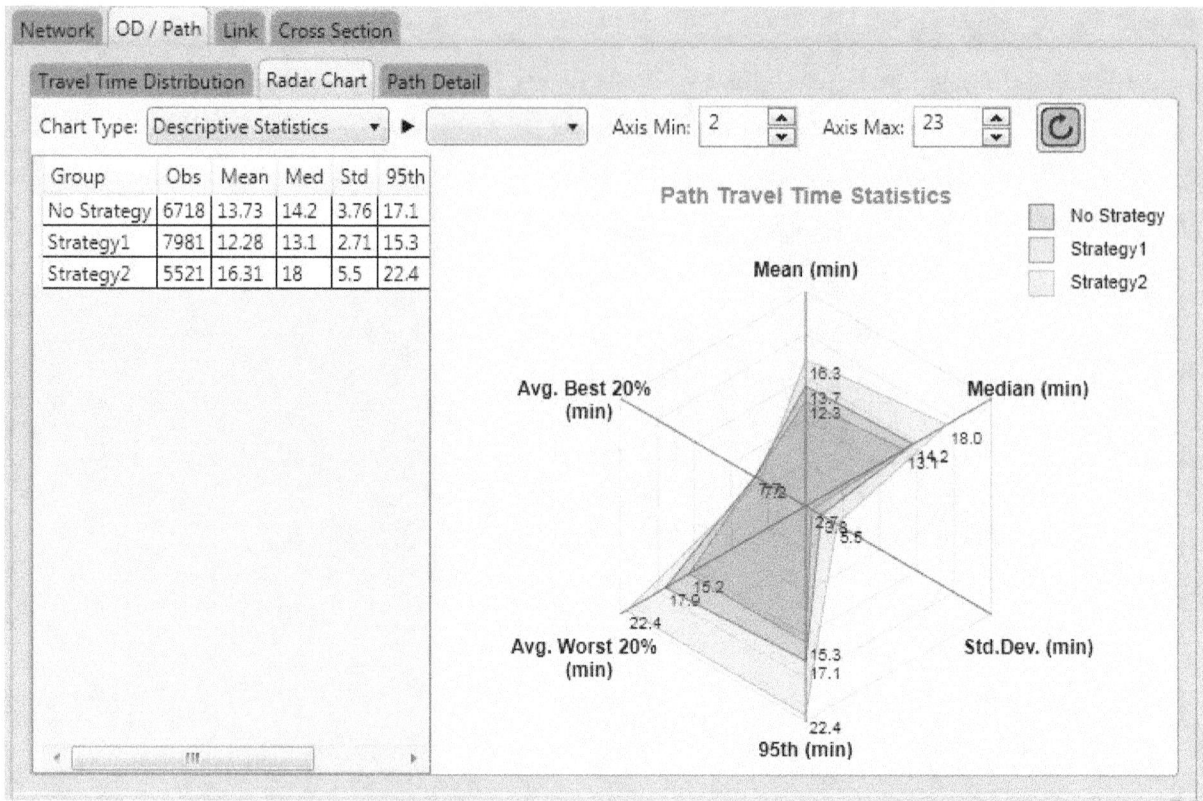

Figure 5-7. Comparing Travel Time Characteristics of Different Scenarios

5.2.2.2 Time-dependent Path Travel Time

While travel time distributions provide rich information on central and dispersion tendencies as discussed above, the temporal dimension of performance measures is missing. In order to examine the temporal evolution of path travel time, one can plot the average path travel time with respect to the departure time interval. The top plot in Figure 5-8 shows the time-dependent average travel times for a selected path, which represent the travel times that are experienced by travelers departing at different aggregate time interval. For example, in Figure 5-8, the average path travel time jumps from 10 minutes to 25 minutes around at the departure time interval 65 minutes under No Strategy (red line); while Strategy 1 maintains the travel time at a stable level between 11 and 15 minutes over time (black line).

5.2.2.3 Identifying Bottleneck Links along Selected Path

In the simulation environment, a path consists of links and the path travel time corresponds to the sum of travel times for those links. By investigating each link travel time separately, one can locate bottleneck links along the path. The bottom plot in Figure 5-8 shows the average link travel time per unit distance for each link along the selected path, where the x-axis represents the sequence of links and each link is expressed as a pair of the upstream and downstream node IDs. The average link travel time per mile (minutes/mile) is used for measuring the link performance, which is the inverse of speed, and links with higher levels of this measure can be considered as bottleneck of the given path. The example in Figure 5-8 shows that the entrance link (Node ID 116 → Node ID 19) is a common bottleneck for all three scenarios; the fourth link (Node ID 28 → Node ID 32) experiences relatively high delay under No Strategy (red) and Strategy 2 (green), but Strategy 1 (black) help resolve the congestion on that link.

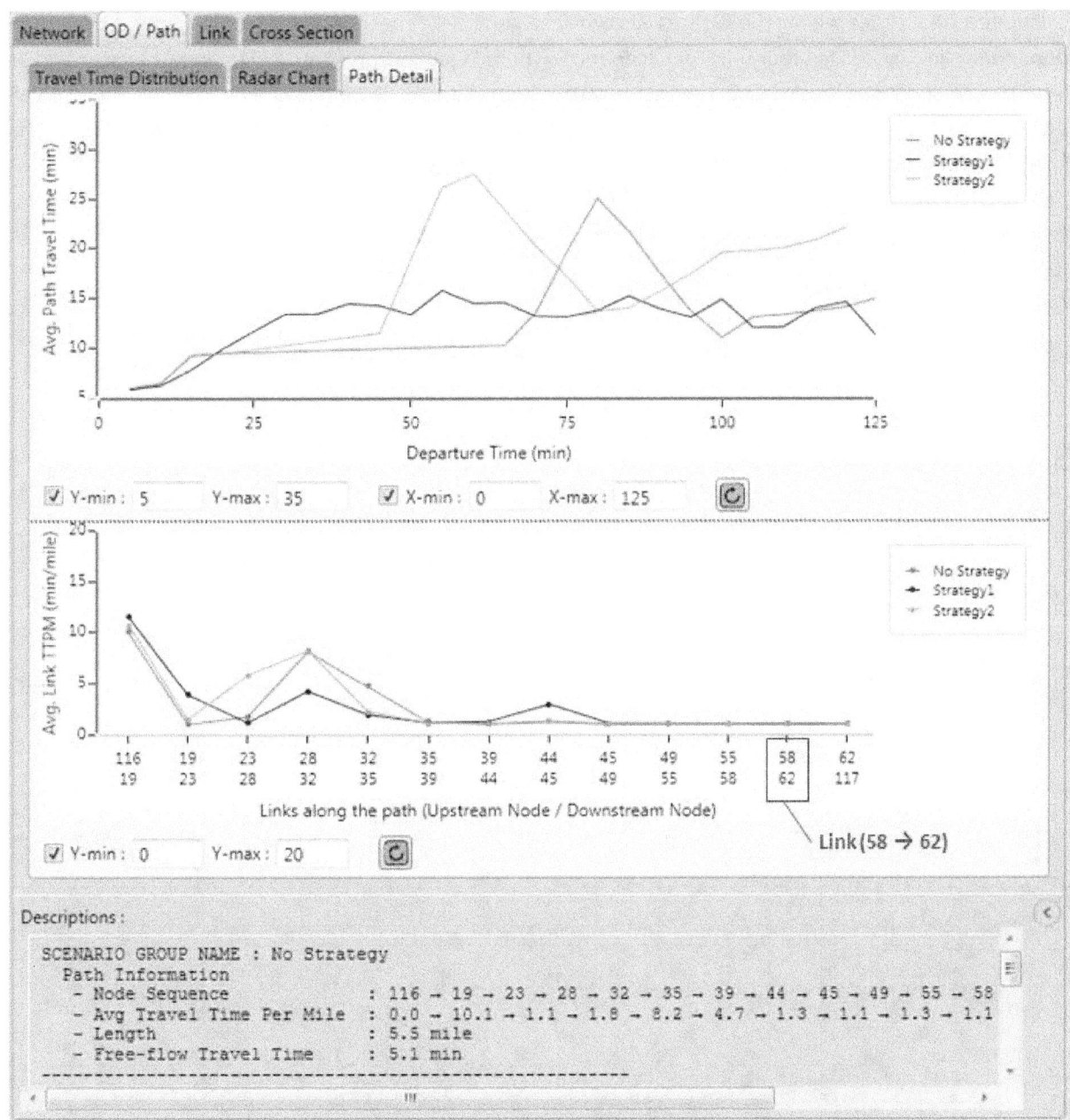

Figure 5-8. Time-dependent Average Path Travel Time (top) and Average Travel Time Per Mile for Each Link along the Path (bottom)

5.2.3 Link-level Measures

For link-level measures, conventional traffic flow parameters such as speed, density and flow rate are calculated and displayed in the GUI as shown in Figure 5-9. The example in this figure plots time-dependent link speeds for four different scenarios, where "Clear" represents a clear weather scenario as a base-case and "Rain" represents the rain weather scenario that decreases the overall link speed level. Two WRTM strategies using variable message signs are tested, where "Rain_VMS1" improves the link speed performance, but "Rain_VMS2" does not.

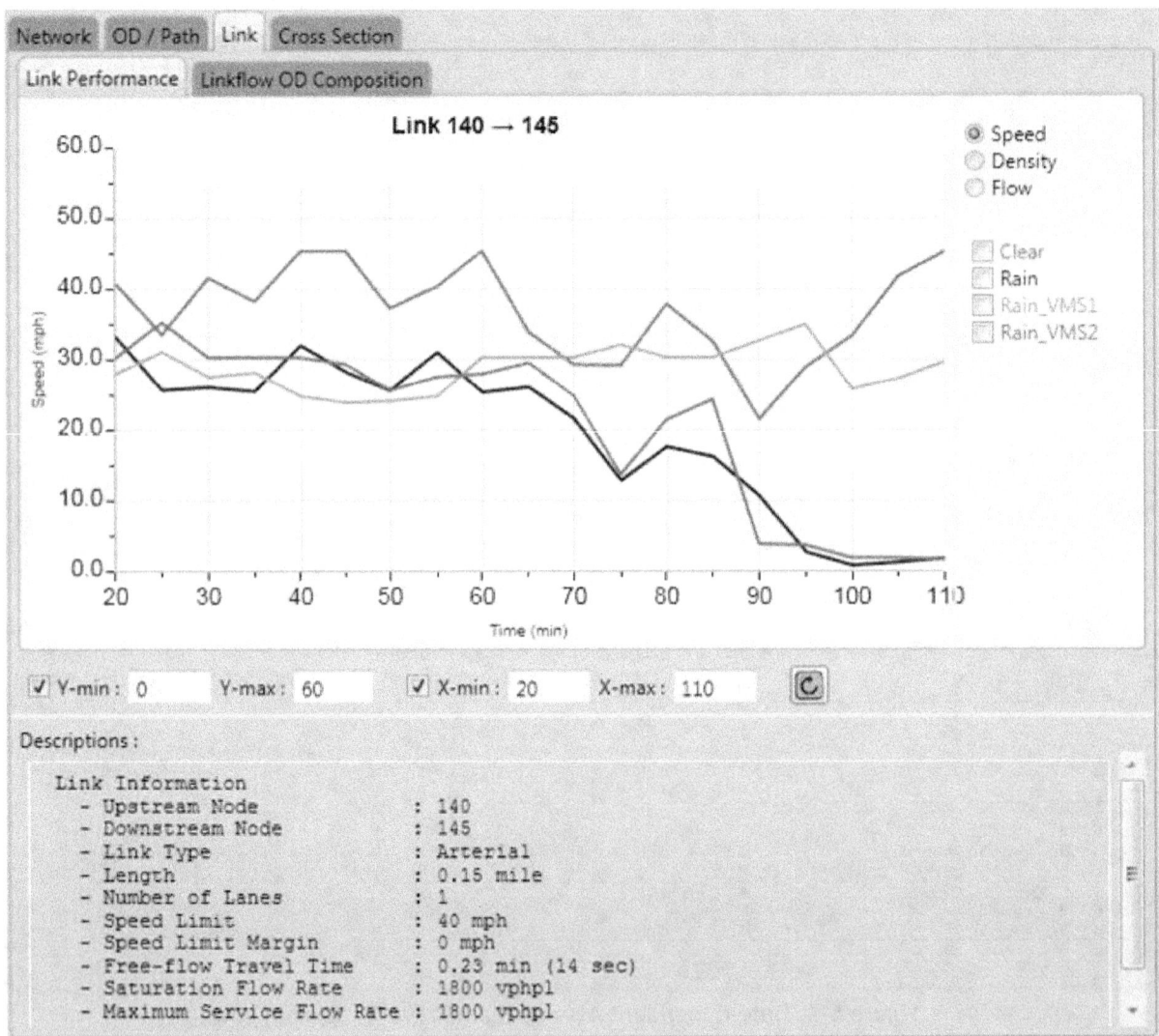

Figure 5-9. Time-dependent Traffic Flow Performance Measures for Selected Link

5.2.4 Cross Section Measures

Another important perspective through which performance measures may be envisioned is a cross-section of a given network. Especially for networks like Long Island, which has clear major flow directions (e.g., eastbound and westbound), TMC operators might be interested in understanding how well the overall traffic flows pass through a certain cross-section under different weather conditions and WRTM strategies. Figure 5-10 shows a GUI that displays a user-defined cross-section on a DYNASMART network and the corresponding Google map for the Long Island area. The vertical bar selects all the westbound links including freeways and arterials. The time-dependent traffic flows aggregated over the selected links are then analyzed as shown in Figure 5-11. The top chart shows the cumulative vehicle counts and the bottom chart shows the dynamic vehicle counts aggregated over every 5 minutes. In this example, two different scenarios are tested, where "base" represents a normal weather condition and "heavy snow" represents a heavy snow weather condition. It can be observed that heavy snow reduces the overall cross-section throughput compared to the clear weather.

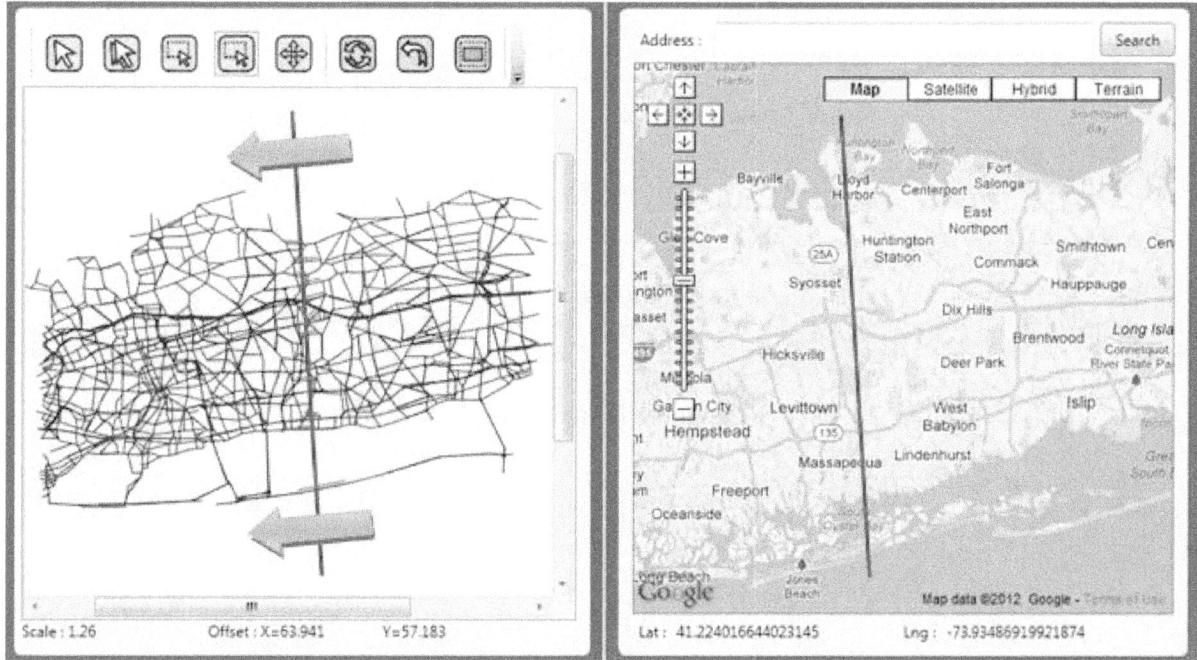

Figure 5-10. Selected Cross Section on Long Island Network (Analysis for Westbound Traffic Flow) (Source: Northwestern University, Trajectory Processor with Google Map API, Accessed May, 2012)

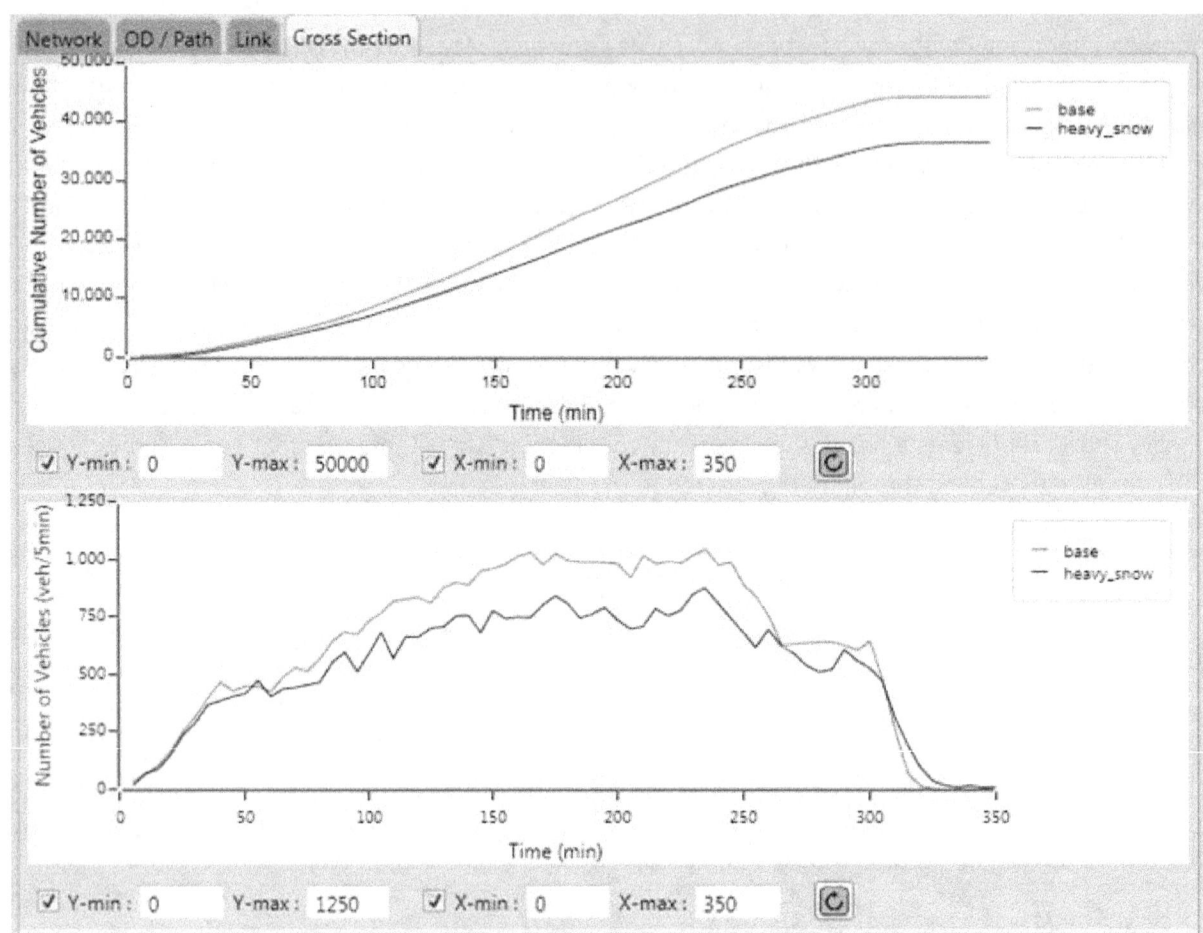

Figure 5-11. Time-dependent Cross-section Throughput Measures

6. Implementation and Evaluation of Selected WRTM Strategies for Study Networks

This chapter discusses the procedures and analysis results for the implementation and evaluation of WRTM strategies conducted for three networks: Chicago, New York (Long Island) and Salt Lake City.

6.1 Chicago Network

6.1.1 Coordination with Chicago DOT

In order to pursue the project tasks in collaboration with the Chicago Department of Transportation (CDOT), a meeting was held on February 21 and 24, 2012 at City of Chicago Department of Transportation in Chicago, IL. The purposes of the meeting include:

- Introduce to the agency the weather-sensitive TrEPS model that allows incorporating weather effects in traffic modeling
- Demonstrate how combining weather forecasts with traffic prediction can help the local agency in weather-related traffic management
- Identify the agency's interest for applying these tool (e.g., on-line implementation, off-line experiments)
- Identify a list of WRTM strategies for the analysis and available data

Based on the discussions and suggestions from the meeting, the Northwestern University team has prepared a sub-network and selected WRTM strategy scenarios to conduct simulation experiments to assess the effectiveness of different strategies under inclement weather conditions.

As CDOT's primary interest is in investigating potential benefits of different WRTM strategies under severe weather conditions and developing new strategies for the future, we focus on performing various what-if scenarios for selected strategies using the planning (off-line) version of the TrEPS model (i.e., DYNASMART-P) in this task. Three strategies are selected: (1) Demand Management, (2) Variable Speed Limit (VSL) and (3) Optional Detour VMS. The first strategy is to address needs for preventing serious deterioration of the network performance under excessively inclement weather conditions like the February 1, 2011, blizzard, in which vehicles were stranded on Lake Shore Drive during a near-record blizzard (a total accumulation of 20.2 inches of snow within hours). The last two strategies are selected based on CDOT's interest in studying the effectiveness of the speed management strategy and real time travel time information on specific routes. These strategies are intended to be used under less extreme weather conditions (e.g., light to moderate snows). We performed various simulation experiments to study the effectiveness of selected strategies.

6.1.2 Sub-network Preparation

6.1.2.1 Suggested Sub-network

CDOT staff and Northwestern University team agreed to extract a smaller network from the entire Chicago Metropolitan Area to enhance the estimation and prediction performance of TrEPS during the implementation procedure. The suggested initial sub-network includes Chicago downtown area located in the central part of the network, and Kennedy Expressway and Edens Expressway. The sub-network is bounded on east by Michigan Lake and on west by Cicero Avenue and Harlem Avenue. Roosevelt Road and Lake Avenue are bounding the sub-network from south and north, respectively. Once the rough boundary of the sub-network is determined, the next step is to prepare all the network-related input files based on the new configuration including a new OD matrix, which reflects zones and travel demand only for the sub-network area. Figure 6-1 depicts the original Chicago network and the extracted sub-network, and Table 6-1 summarizes characteristics of the two networks.

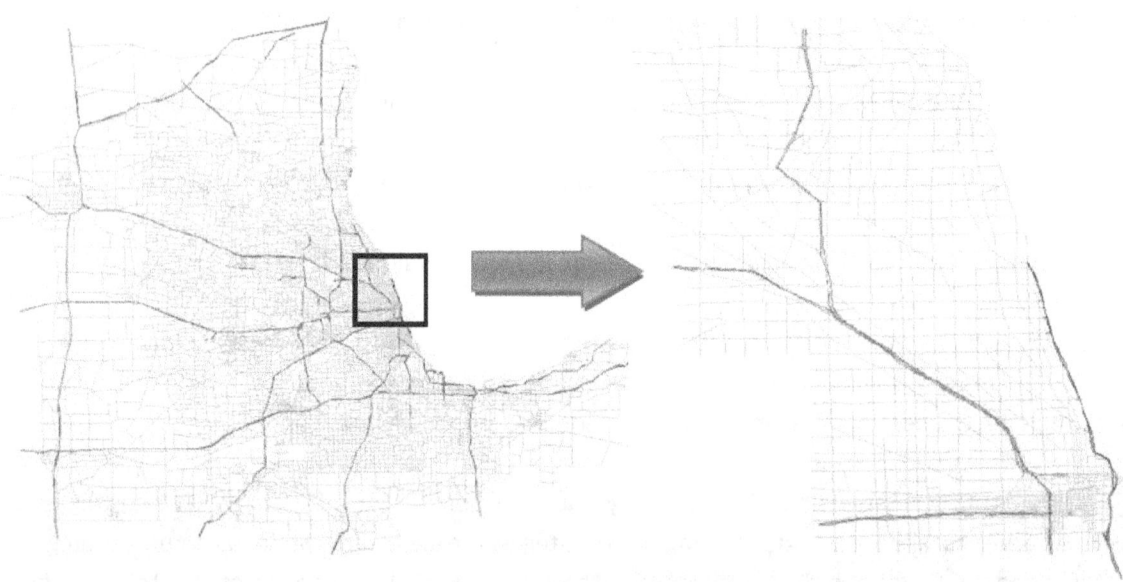

Figure 6-1. Map of the Extracted Network of Chicago

Table 6-1. Comparing Network Characteristics for Original and Extracted Networks of Chicago

Network	Original Chicago Network	Chicago Sub-network

Description	<ul><li>40,443 links
	   144 links are tolled
	   1,400 freeways
	   201 highways
	   2,120 ramps (96 of them are metered)
	   36,722 arterials</li><li>13,093 nodes
	   2155 signalized intersections
	   1,961 zones
	   1,944 internal
	   17 external</li><li>Demand period
	   5am -10am hourly demand (~4,100,000 total demand)</li></ul>
		<ul><li>4,805 links
		   No tolled links
		   150 freeways
		   47 highways
		   247 ramps (59 of them are metered)
		   4,361 arterials</li><li>1,578 nodes
		   545 signalized intersections
		   218 zones</li><li>Demand period
		   5am -11am hourly demand (~800,000 total demand)</li></ul>

6.1.2.2 Procedures for Network Extraction and OD Estimation

The network extraction consists of the following three major steps:

Step 1. A new subnetwork is defined by the sets of nodes N, links A and zones Z according to the new boundaries decided by the planners. This subnetwork is designated as the internalnetwork, whereas the remaing sections of the network is designated as the externalnetwork.

Step 2. The original (external+internal) network is simulated in a dynamic simulation and assignment platform (e.g. DYNASMART-P) to obtain the following flows:
   a. External to internal,
   b. Internal to external,
   c. Internal to internal,
   d. External to external (using the internal network),
   e. External to external (not using the internal network).

The sum of these 5 flows defines the time-dependent origin-destination (TDOD) matrix of the original network, whereas the sum of the first 4 flows defines the TDOD matrix of the subnetwork of interest. The outcome of this step is the TDOD matrix of the subnetwork.

These steps make up the extraction of the subnetwork with the sets of nodes N, links A, zones Z and the TDOD matrix. Figure 6-2 shows the time-dependent profile of extracted travel demand for the sub-network.

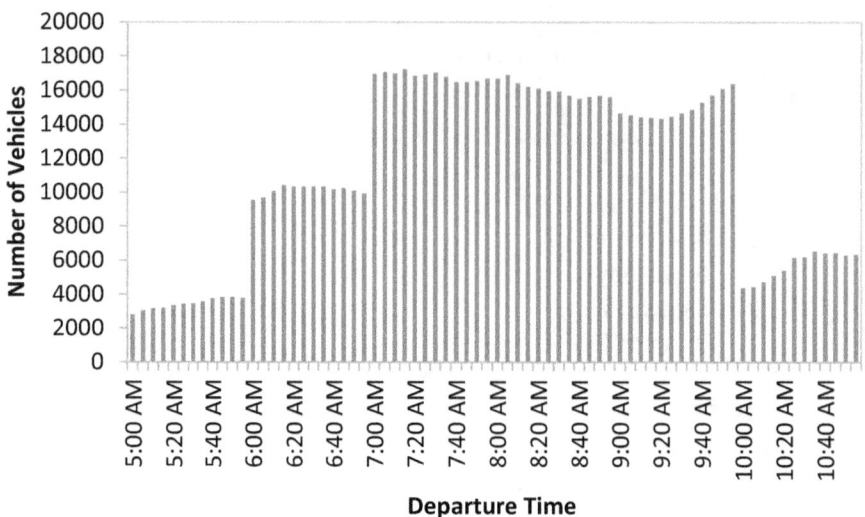

Figure 6-2. Temporal Distribution of Demand for Sub-network from 5:00 AM to 11:00 AM

6.1.3 Implementation and Evaluation of WRTM Strategies: (1) Demand Management

6.1.3.1 Weather-responsive Demand Management

Managing demand in this study is about providing travelers with information, aiming at a "shift" of their departure times or trip cancelation so that the total travel demand during the peak periods can be reduced. The key research question here is to study how much demand should be reduced under different weather conditions in order to maintain a certain level of network performance. It is critical for the TMC operators to provide "reliable" information to maintain credibility with roadway users. It is also important to try to minimize the potential economic losses by setting the target demand to its necessary level. Attempting to reduce demand beyond this level might cause significant financial loss to the local business and community. As such, the goal of using TrEPS here is to provide TMC operators with the information on the optimal level of demand that can improve the network performance but not affect negatively the productivity under a given weather condition.

6.1.3.2 Experiment Design

Weather Scenario

As the initial motivation for the demand management is the February 1, 2011, blizzard, which was the third largest blizzard in Chicago history and lead to CDOT's keen interest in exploring possible response activities to assist in decision making on future snow incidents, we construct a heavy snow scenario based on the historical data collected on February 1, 2011 in Chicago. Figure 6-3 depicts the profiles of the snow (liquid equivalent) precipitation (inch/hour) and the visibility (mile) for the generated snow scenario, which is extracted from the time period between 12:45PM and 8:45PM. This 8-hour weather scenario is used for the simulation of the sub-network. The entire simulation horizon is 8 hours, where vehicles are generated and loaded into the network during the first 6 hours based on the OD matrix, which represents the traffic demand between 5:00 AM to 11:00 AM. For the remaining 2 hours, vehicles are simply simulated so that the generated vehicles reach their destinations.

Demand Scenarios

Total 12 demand scenarios are prepared: one for the benchmark case, which is 100% of the demand under the normal weather condition (i.e., no snow); and the other 11 scenarios with different demand levels under the heavy snow condition. For the generation of the 11 scenarios, we start with the full demand (100%) and reduce the total demand by 5% until the reduction percentage reaches 50%. The purpose of this experiment is to answer the question: "how much reduction in overall demand should we try to achieve under a particular weather scenario in order to maintain the same level of service on our network"?

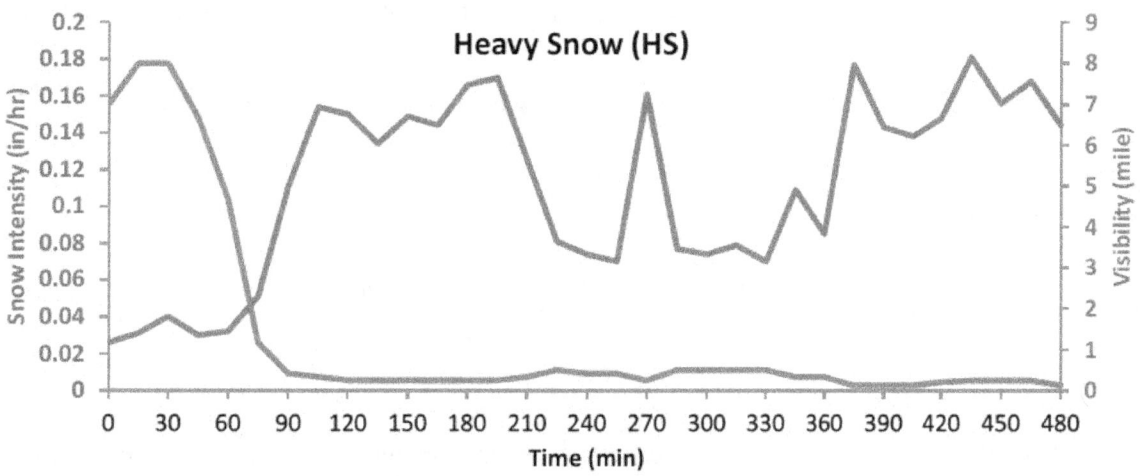

Figure 6-3. Weather Scenario for Demand Management Strategy:
Heavy Snow from Historical Data (extracted from 2011-02-01 12:45PM – 8:45PM)

6.1.3.3 Analysis Results

Performance Measures

A set of network performance measures are defined to illustrate network-level traffic conditions under different weather and demand scenarios.

- Accumulated Percentage of Out-Vehicle

The accumulated percentage of out-vehicle is the percentage of vehicles arriving at their destinations from the start of the simulation till a given time stamp t. It can be expressed in the following form:

$$\%Accumulated\_Out\_Veh_i^t = \frac{Out\_Veh_i^t}{Tot\_Veh_i^t} \times 100 \qquad (6\text{-}1)$$

Where

$Out\_Veh_i^t$ — Accumulated number of vehicles arriving their destinations from time 0 till time *t* in scenario *i*

$Tot\_Veh_i^t$ — Accumulated total number of vehicles loaded onto the network from time 0 till time *t* in scenario *i*

- *Percentage Change in Average Travel Time*

$$\%change\_AvgTTime_i^k = \frac{\Delta AvgTTime_i^k}{AvgTTime_{base}} = \frac{AvgTTime_i^k - AvgTTime_{base}}{AvgTTime_{base}} \times 100 \quad (6\text{-}2)$$

Where

$AvgTTime_{base}$ — Average travel time for full demand in the base case without weather feature

$AvgTTime_i^k$ — Average travel time for *k* percent of full demand in weather scenario *i*

- *Percentage Change in Average Stop Time*

$$\%change\_AvgSTime_i^k = \frac{\Delta AvgSTime_i^k}{AvgSTime_{base}} = \frac{AvgSTime_i^k - AvgSTime_{base}}{AvgSTime_{base}} \times 100 \quad (6\text{-}3)$$

Where

$AvgSTime_{base}$ — Average travel time for full demand in the base case without weather feature

$AvgSTime_i^k$ — Average travel time for *k* percent of full demand in weather scenario *i*

Discussion

Figure 6-4 shows the accumulated percentage of out-vehicles representing throughput of the network under different scenarios. One might notice that there are several points where sudden drops or slight jumps are observed in the chart. This is due to the time-dependent demand pattern shown in Figure 6-2 and those points correspond to the time points in which the demand increases or decreases. In addition to the demand, the weather profile also affects the throughput pattern because the snow intensity and visibility values change over time as shown in Figure 6-3 and this change reduces or increases the supply-side capacity of the network.

Compared to the benchmark case (i.e., Benchmark), where no snow event is present, the snow effect significantly deteriorates the network throughput if the original full demand is used, i.e., no demand management is applied (i.e., Heavy Snow (100% Demand)). It can be seen that the network throughput decreases by about 10% due to weather. It is observed, however, the network performance gets better as the demand level decreases. By reducing the demand by 20 to 25% (i.e., Heavy Snow (80% Demand) or Heavy Snow (75% Demand)), we can achieve the network throughput similar to the original level (i.e., Benchmark) at the end of the simulation.

Figure 6-5 presents the percentage changes in the average travel time and the average stop time for different demand scenarios relative to the benchmark case. With 25% percent of demand reduction, both measures are recovered to the level of the benchmark case.

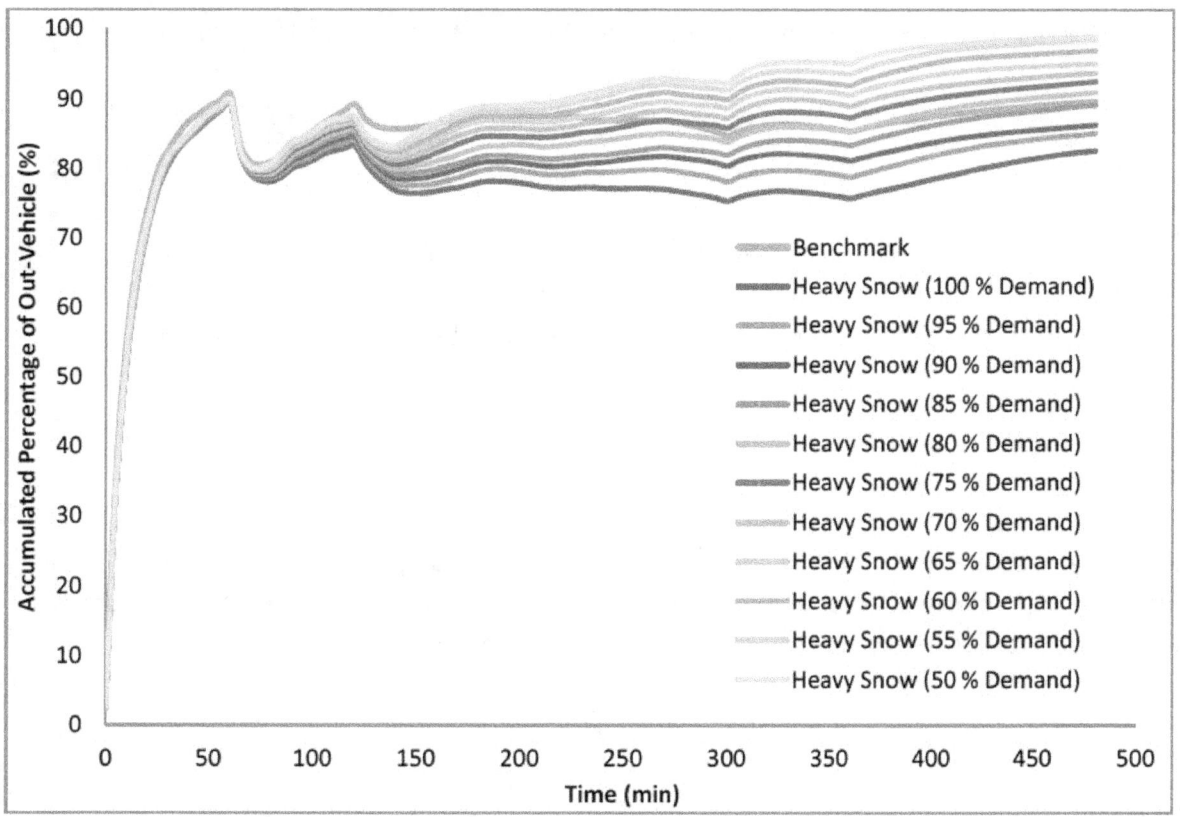

Figure 6-4. Accumulated Percentage of Out-Vehicle for Different Scenarios

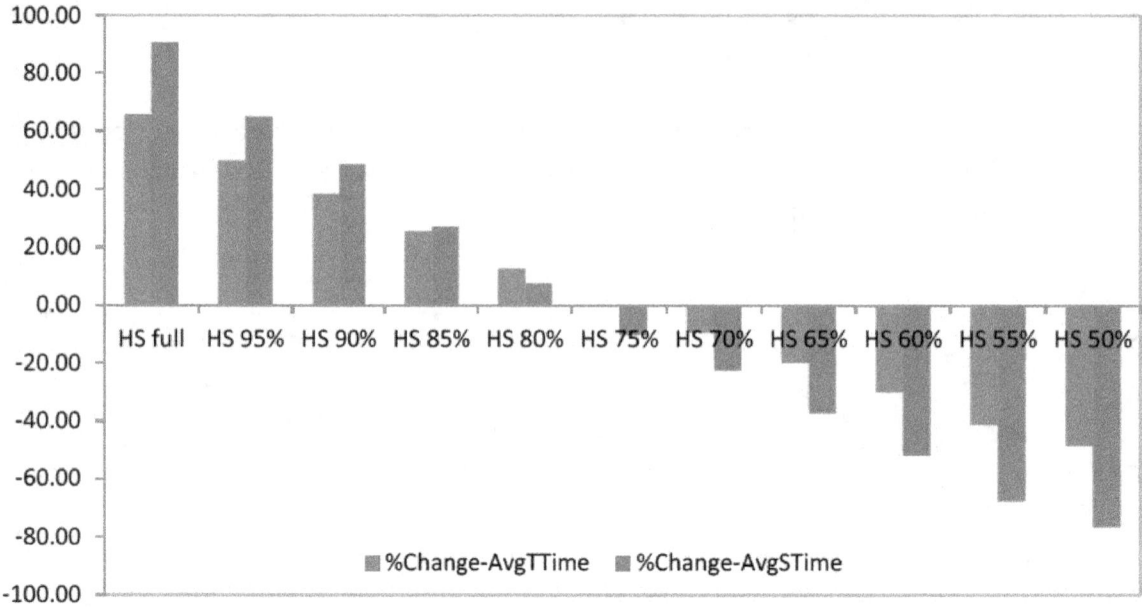

Figure 6-5. Changes in Average Travel Time and Average Stop Time Relative to Benchmark

6.1.4 Implementation and Evaluation of WRTM Strategies: (2) Variable Speed Limit

6.1.4.1 Weather-responsive Variable Speed Limit (VSL)

The weather-responsive VSL strategy is a traffic management strategy that utilizes weather information to determine appropriate speeds at which drivers should be traveling, given current roadway conditions. These advisory or regulatory speeds are usually displayed on overhead or roadside variable message signs (VMS).

6.1.4.2 Experiment Design

Weather Scenario

For the implementation of the VSL strategy, we construct a moderate snow scenario based on the historical data collected on December 12th 2010 from 10:00 AM to 6:00 PM in Chicago. Figure 6-6 depicts temporal profiles of the snow (liquid equivalent) precipitation intensity (inch/hour) and the visibility (mile) of Moderate Snow (MS) scenario. This 8-hour weather scenario is used for the simulation of the sub-network. The entire simulation horizon is 8 hours, where the first 6 hours are used for generating and moving vehicles based on the estimated OD matrix, which represents the traffic demand between 5:00 AM to 11:00 AM, and the remaining 2 hours are used for simply moving generated vehicles to their destinations.

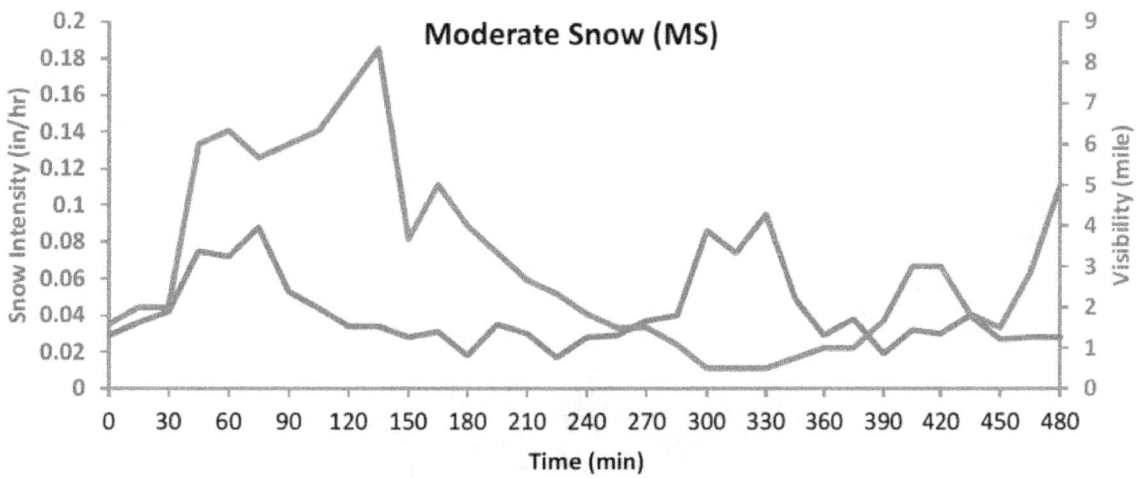

Figure 6-6. Weather Scenario for VSL Strategy:
Moderate Snow from Historical Data (extracted from 2010-12-12 10:00AM – 6:00PM)

Variable Speed Limit Scenarios

Two variables are considered in constructing the VSL scenarios: the location of VSL signs and the start time of the VSL implementation. For the VSL location, we examine three different cases: applying VSL on both freeways and Lake Shore Drive (Figure 6-7 (a) Case 1); applying VSL on Lake Shore Drive only(Figure 6-7 (b) Case 2); and applying VSL on a selected section on Kennedy Expressway only (Figure 6-7 (c) Case 3). In Case 1, applying variable speed limit on all the major corridors might not seem realistic, but we included the results here for the comparison. Based on the initial observation on the overall congestion level under the snow scenario, we found that Lake Shore Drive and Kennedy Expressway are the most congested corridors and applied VSL focusing on those two roads in Case 2 and 3. In Case 3, to select a particular section on Kennedy Expressway for the implementation of VSL, we first examined the traffic state along the freeway under the given weather scenario. Then the most congested section, where the initial traffic queues form and propagate upstream, is selected for the VSL operations. This is based on the assumption that the VSL has the effect of preventing or delaying the onset of flow breakdown by harmonizing the link speed as well as leading vehicles to other routes so that the congestion on the given section could be relieved and the overall network performance could benefit from it. In addition to these three cases, we have tried many other different scenarios, but the results are not presented here because those are found less effective than the above-mentioned three cases.

For the start time choice of the VSL implementation, we analyzed various starting points from 30 minutes to 180 minutes by incrementing 30minutes. The final scenarios are selected as 90, 120, 150 and 180 minutes for the VSL activation. Table 6-2 summarizes all the 12 scenarios constructed by the VSL location and start time choices.

Table 6-2 Constructing Variable Speed Limit Strategy Scenarios

		VSL Start-Time (minutes)			
		90	120	150	180
VSL Locations	Case 1	√	√	√	√
	Case 2	√	√	√	√
	Case 3	√	√	√	√

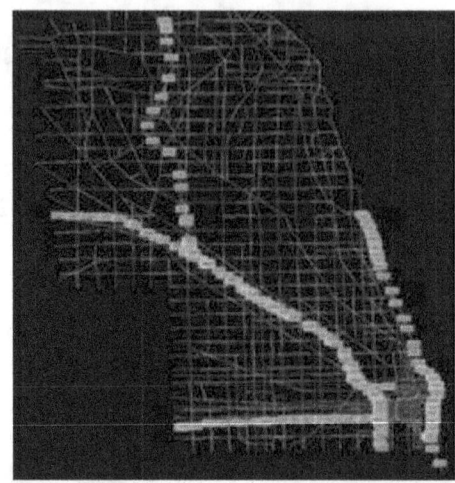

a) Case 1: VSL on Both Freeways and Lake Shore Drive

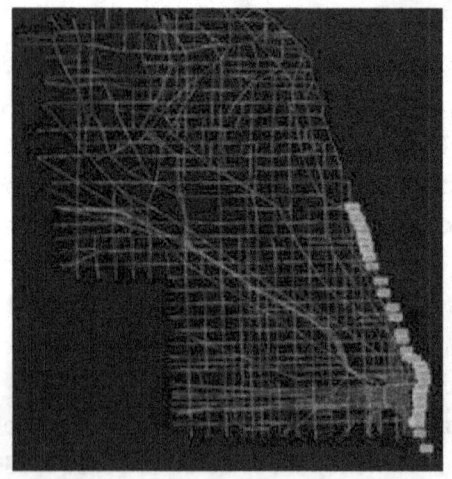 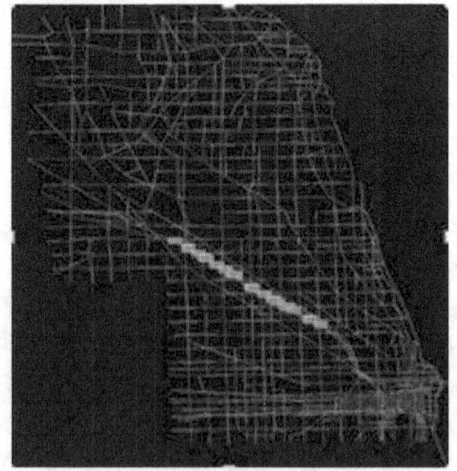

b) Case2: VSL on Lake Shore Drive Only   c) Case3: VSL on a selected section on Kennedy Expressway

Figure 6-7. Location of Variable Speed Limit Signs for Different Scenarios

6.1.4.3 Analysis Results

Initial Evaluation

As an initial evaluation, two performance measures are selected to grasp the overall effectiveness of different VSL strategies: (1) the number of vehicles left in the network at the end of simulation horizon and (2) total travel time (i.e., the sum of the travel times for all vehicles). Figure 6-8 and Figure 6-9 show the former and the latter for the different scenarios, respectively. In the both figures, the benchmark case (i.e., No Snow and No VSL) and the no-VSL case (i.e., Snow, No VSL) are also presented for the comparison. Note that vehicles still in the network are those that have been generated from their respective origins at some point during the simulation period but have not yet reached their destination at the end of the simulation horizon. This number generally increases during weather events as the overall throughput decreases and traversal times correspondingly increase.

The resulting measures vary greatly with the VSL locations and the start times. Case 2 (Lake Shore Dr) exhibits the best performance and Case 3 (Kennedy Expy) the worst performance when the VSL starts at 90 minutes, while Case 2 exhibits the worst and Case 3 the best when it starts at 120 minutes.

In terms of the actual benefit of the strategy under the snow event, which can be measured by the performance difference between "Snow+No VSL" and "Snow+VSL" cases, there are three scenarios that show such improvements in both the number of vehicles left and the total travel time:

- VSL (Case 2, Lake Shore Dr) starting at minute 90 (*the number of vehicles left is reduced by 4,530 and the total travel time is reduced by 10,800 hours*);
- VSL (Case 3, Kennedy Expy) starting at minute 120 (*the number of vehicles left is reduced by 3,990 and the total travel time is reduced by 13,300 hours*); and
- VSL (Case1, both freeway and Lake Shore Dr) starting at minute 120 (*the number of vehicles left is reduced by 280 and the total travel time is reduced by 7,700 hours*).

One important point from the initial results of the experiment is that the effectiveness of the VSL strategy highly depends on its start-time choice. For instance, applying VSL too early may unnecessarily slow down the traffic and lead to negative effects; while applying too late would not help preventing congestion as it is too late to intervene. Furthermore, applying VSL on the congested part of the freeway instead of entire freeway network has better performance as a result of not slowing down vehicles in the uncongested parts of the network. It appears that this is where the on-line TrEPS comes into play. By performing the simulation analysis in real-time through the TrEPS, TMC operators could dynamically determine the appropriate timings for turning on and off the VSL signs based on the prevailing and predicted traffic states.

The analysis results presented here are intended to provide an indication regarding how the TrEPS could be used in implementing and evaluating the VSL strategies, rather than to identify a clear pattern or draw a definitive conclusion about the impact of the VSL strategies as the analysis results are also dependent on the input weather scenarios. It is also worth noting that another important benefit of applying the VSL is on the safety side, which may not be reflected in the mobility-based performance measures considered in the simulation model output.

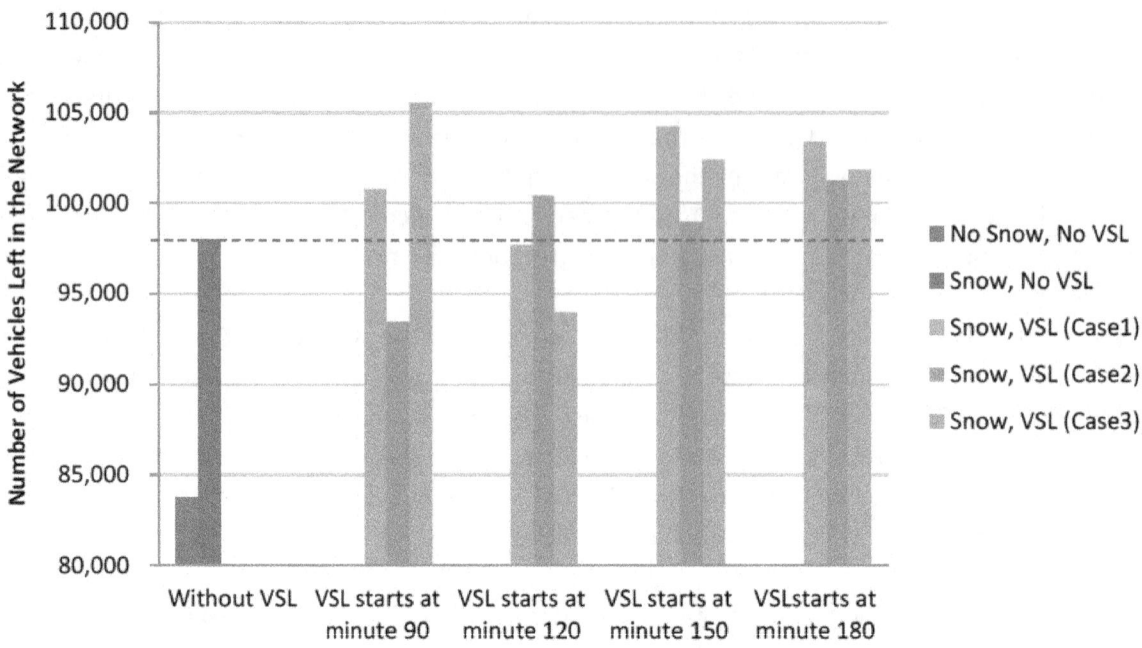

Figure 6-8. Number of Vehicles Remaining in the Network at the End of Simulation

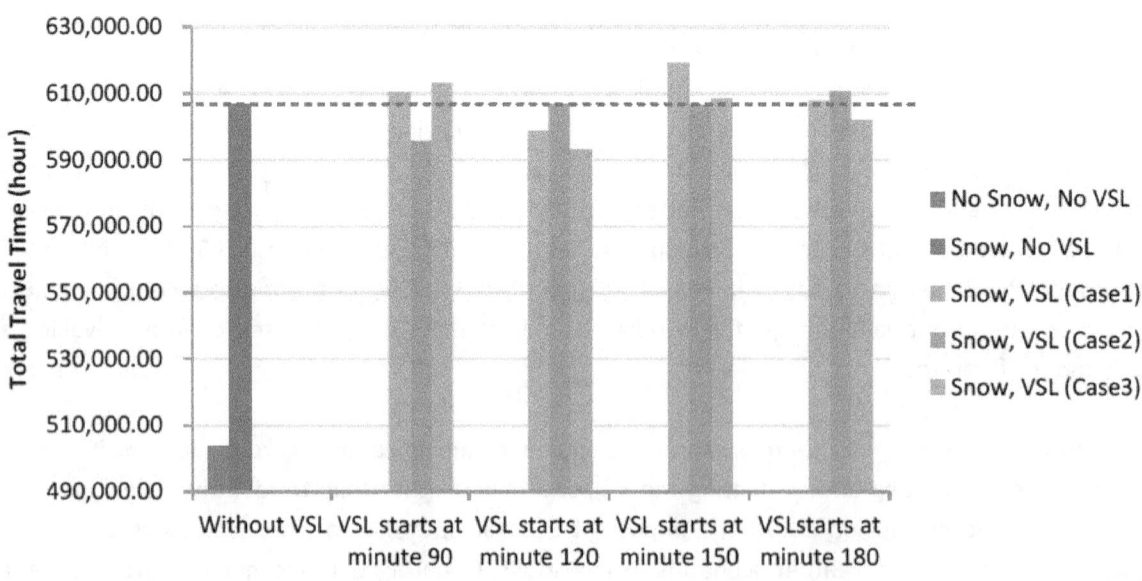

Figure 6-9. Total Travel Time in Hours for Different VSL Scenarios

Detailed Evaluation

Based on the initial evaluation results, four scenarios are selected for a more detailed evaluation. The objective of the detailed evaluation is to better understand the benefits of the WRTM strategies that showed overall good performance in the initial evaluation by performing an in-depth investigation using more diverse performance measures. The four selected scenarios are presented in Table 6-3, where new scenario names are assigned: "Base" represents the base-case scenario with no snow and no VSL; "SN" represents the snow-only scenario; "SN_VSL1" corresponds to the VSL strategy that showed the best performance among those started at minute 90 (i.e., Case 2: Lake Shore Dr); and "SN_VSL2" corresponds to the VSL strategy that showed the best performance among those started at minute 120 (i.e., Case 3: Kennedy Expy).

Table 6-3. Selected Scenarios for Detailed Evaluation

Scenario	Description		
	Snow	VSL	VMS Strategy
Base	No	No	-
SN	Yes	No	-
SN_VSL1	Yes	Yes	(Case 2) Lake Shore Drive at Minute 90
SN_VSL2	Yes	Yes	(Case 3) Kennedy Expy at Minute 120

With these four scenarios, we observe traffic throughput for a selected cross-section to understand how the aggregated vehicle flows over the cross-section change under different strategies. Figure 6-10 shows the selected cross-section on a DYNASMART network and the corresponding Google map for the Chicago sub-network. The horizontal bar selects all the northbound links including freeways and arterials. The time-dependent traffic flows aggregated over the selected links are then analyzed. Figure 6-11presents the cumulative vehicle counts measured for those links. "Base" and "SN" show the best and worst performances, respectively. It is observed that "SN_VSL2" improves the cross-section throughput under the snow condition compared to the no-strategy case (i.e., "SN"). The benefit of the first VSL strategy "SN_VSL1" is very slight.

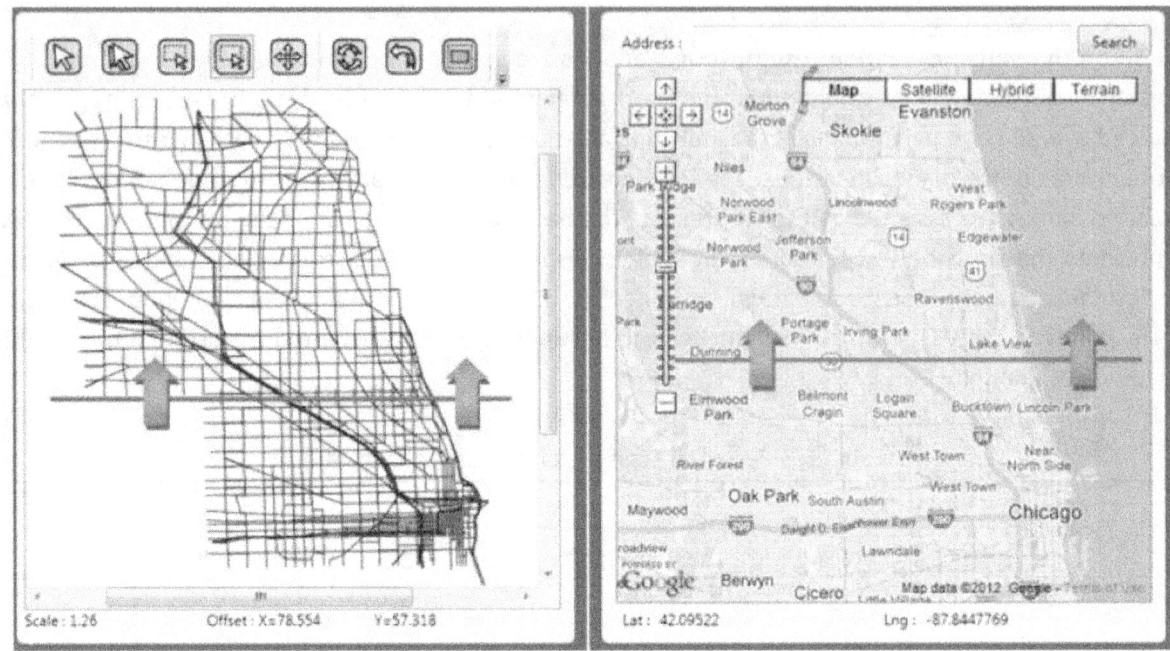

Figure 6-10. Selected Cross Section for Measuring Traffic Throughput (Northbound)
(Source: Northwestern University, Trajectory Processor with Google Map API, Accessed May, 2012)

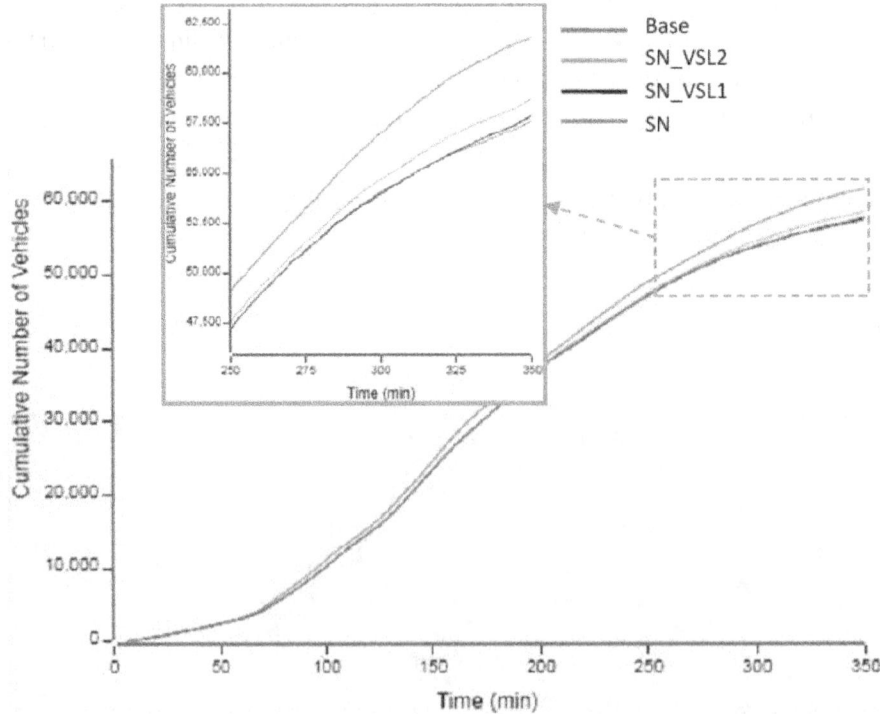

Figure 6-11. Time-dependent Cross-section Throughput Measures (Cumulative Flows)

6.1.5 Implementation and Evaluation of WRTM Strategies: (3) Optional Detour VMS

6.1.5.1 Weather-responsive Variable Message Signs (Optional Detour)

The Optional Detour VMS represents the variable message signs that display the roadway information (e.g., the traffic congestion ahead) as well as possible detour paths so that drivers could reevaluate their routes and divert if a better path exists.

6.1.5.2 Experiment Design

Weather Scenario

For the implementation of the Optional Detour VMS strategy, we used the same weather scenario as the VSL testing as shown in Figure 6-6.

Optional Detour Scenarios

Two variables are considered in constructing the Optional Detour VMS scenarios: the location of detour signs and the percentage of responsive users to these signs. The percentage of responsive users is the percentage of drivers who will re-evaluate their travel time using detour information and makes their decision to get or not to get the detour path. For the detour location, we examined three different cases: Detour option at each exit along the selected freeway section and Lake Shore Drive (Figure 6-12 (a)); Detour option at the end of Lake Shore Drive only(Figure 6-12(b)); and Detour option along Kennedy Expressway only (Figure 6-12(c)). In addition to the spatial distribution, we also considered two different percentages of responsive users: 50 % and 100 %. The 50% response rate indicates that 50 percent of all vehicles generated in the network will respond to a detour sign if they observe it along their respective paths. As a result, a total of six scenarios are used for the simulation study.

6.1.5.3 Analysis Results

Initial Evaluation

Similar to the VSL discussed in the previous section, two performance measures are selected for the initial evaluation to grasp the overall effectiveness of different Optional Detour VMS strategies: (1) the number of vehicles left in the network at the end of simulation and (2) total travel time. Figure 6-13 and Figure 6-14show the former and the latter for the different scenarios, respectively. In both figures, the benchmark case and the no-Detour case are also presented for comparison.

The Optional Detour VMS strategies result in considerable improvement in the performance measures in most scenarios, except one case (i.e., Case 2 with 50 % of responsive users). In Figure 6-13and Figure 6-14, we can observe both the number of vehicles left in the network and the total travel times are reduced when applying the optional detour VMS strategies compared to the case where no strategy is used (i.e., Snow, No VMS). In terms of the number of vehicles left in the network, these VMS strategies produce even better performance than that from the normal weather scenario (i.e., No Snow, No VMS). For the different VMS locations, it is found that Case 1 (applying the optional detour VMS to both freeway and Lake Shore Drive) performs the best, followed by Case 3 (detour VMS on the freeway section only). One interesting observation is found in Case 2 (detour VMS on Lake Shore Drive only), which shows similar or even better results than other cases do under the 100% response rate, but shows significant performance drops under the 50% response rate.

In terms of the user response rate, when we assume a high percentage of users (100%), who responds the optional detour VMS, we observe better performance measures than when we assume a low percentage (50%) of responsive users. This indicates the importance of the real-time roadway information provided to users, which can be effectively used for redistributing and balancing traffic flows thereby preventing serious congestion.

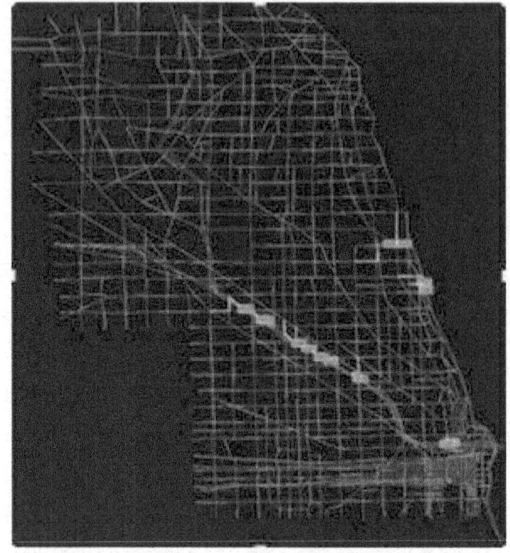

a) Case 1: Detour option along the freeway section and Lake Shore Drive

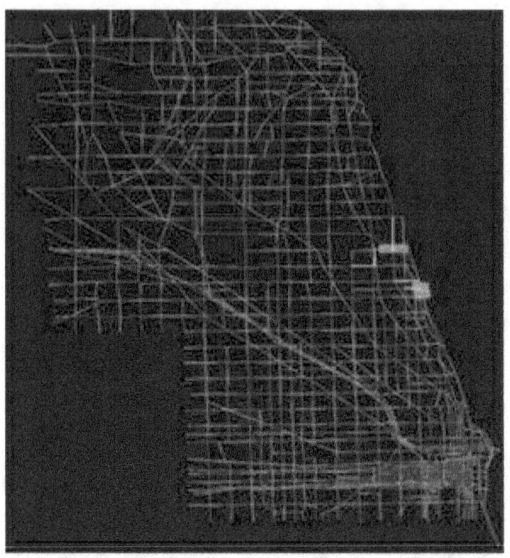

b) Case 2: Detour option at the end of Lake Shore Drive

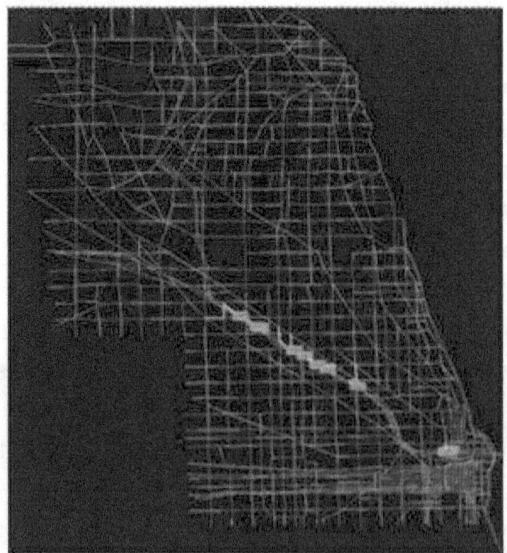

c) Case 3: Detour option along Kennedy Expressway

Figure 6-12. Location of Detour Signs for Different Scenarios

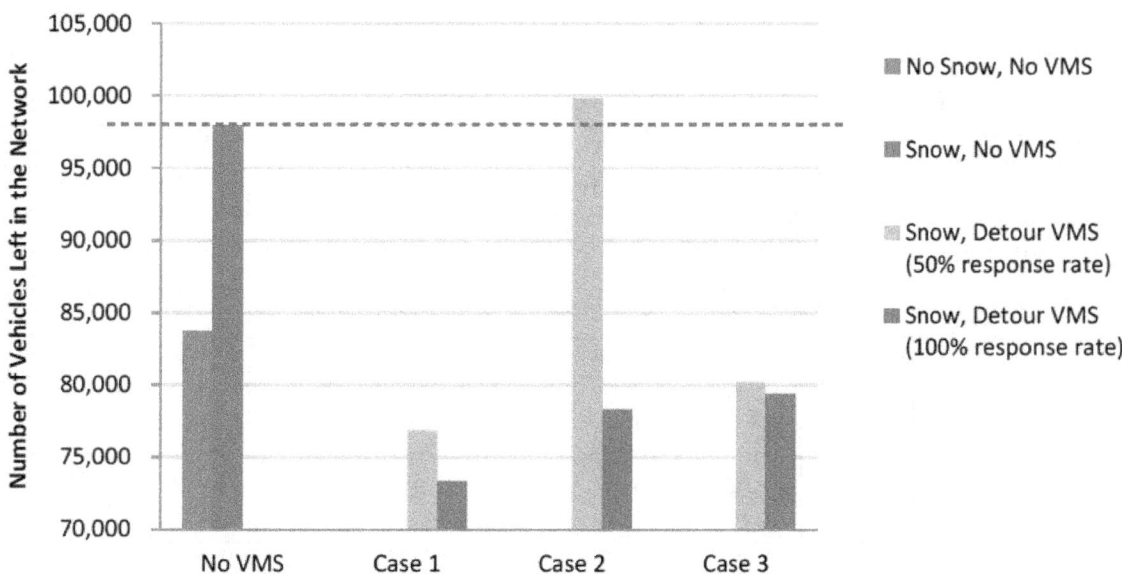

Figure 6-13. Number of Vehicles Remaining in the Network at the End of Simulation

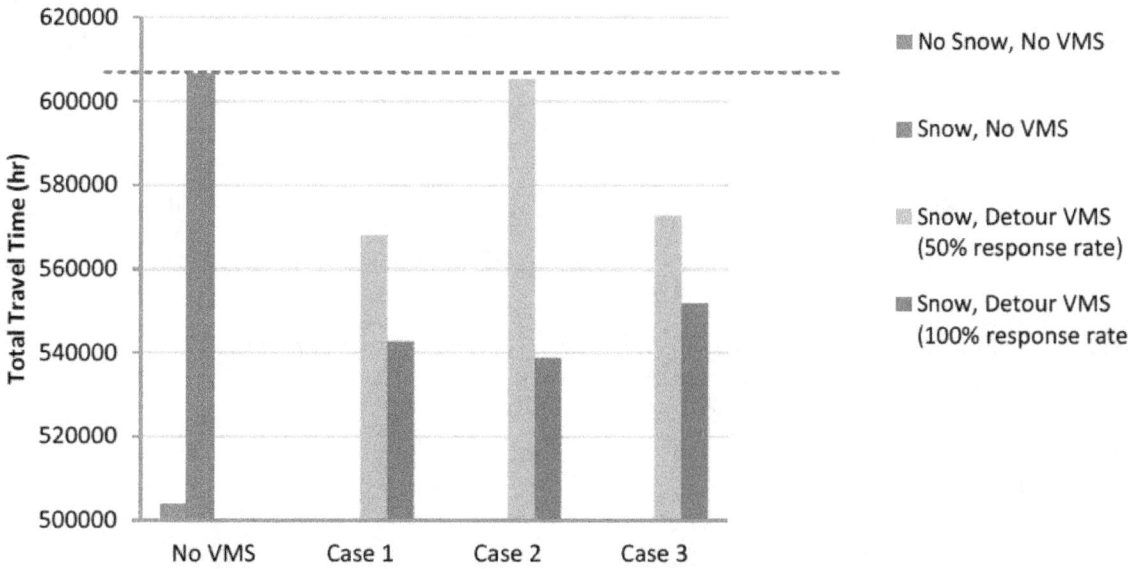

Figure 6-14. Total Travel Time in Hours for Different Optional Detour VMS Scenarios

Detailed Evaluation

Based on the initial evaluation results, four scenarios are selected for a more detailed evaluation. The objective of the detailed evaluation is to better understand the benefits of the WRTM strategies that showed overall good performance in the initial evaluation by performing an in-depth investigation using more diverse performance measrues. In addition to the four selected scenarios in Table 6-3, two more

VMS scenarios are included in Table 6-4, where new scenario names are also assigned: "SN_VMS1" corresponds to the Optional Detour VMS strategy that showed the best performance among those with 50% response rate (i.e., Case 1); and "SN_VMS2" corresponds to the Optional Detour VMS strategy that showed the second best performance among those with 50% response rate (i.e., Case 3).

Table 6-4. Selected Scenarios for Detailed Evaluation

Scenario	Description		
	Snow	VSL	VMS Strategy
Base	No	No	-
SN	Yes	No	-
SN_VSL1	Yes	Yes	(Case 2 in Figure 6-7) Lake Shore Drive at Minute 90
SN_VSL2	Yes	Yes	(Case 3 in Figure 6-7) Kennedy Expy at Minute 120
SN_VMS1	**Yes**	**Yes**	(Case 1 in Figure 6-12) Lake Shore Drive and freeways + 50% response rate
SN_VMS2	**Yes**	**Yes**	(Case 3 in Figure 6-12) Kennedy Expy only + 50% response rate

With these all six scenarios, we again observe traffic throughput for the selected cross-section to understand how the aggregated vehicle flows over the cross-section change under different strategies. The same section is used for the analysis as in Figure 6-10. Figure 6-15 presents the cumulative vehicle counts measured for the relevant links. Overall, the optional detour VMS strategies perform better than VSL strategies as both two scenarios "SN_VMS1" and "SN_VMS2" improve the cross-section throughput under the snow condition.

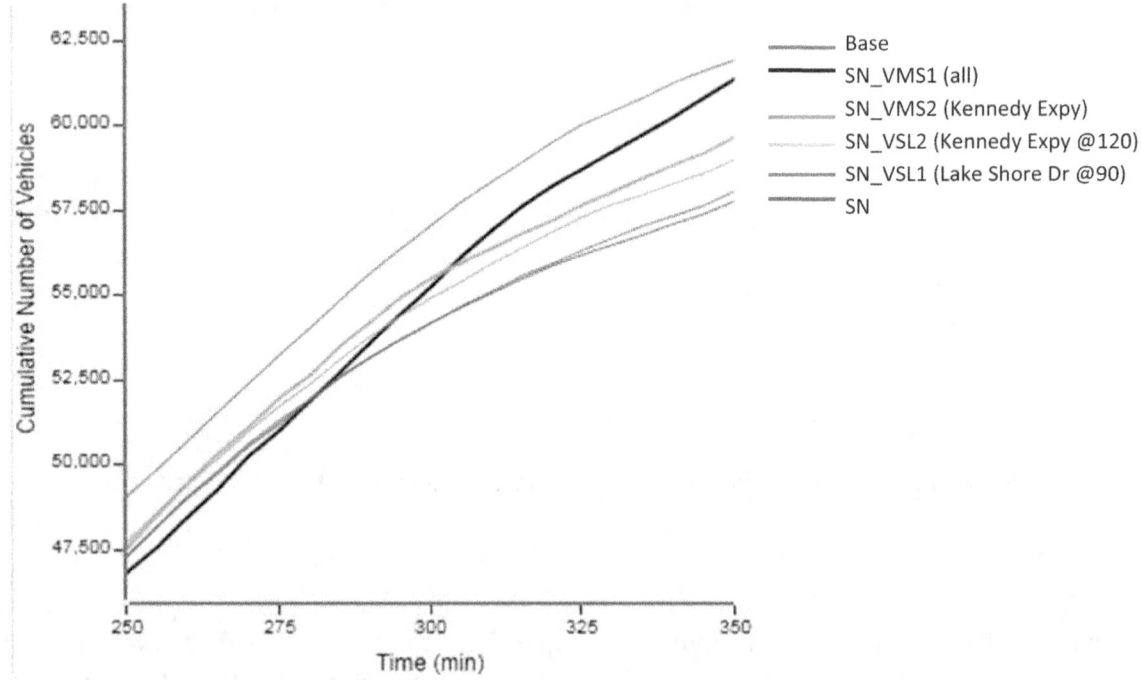

Figure 6-15. Time-dependent Cross-section Throughput Measures (Cumulative Flows)

6.2 Long Island Network

6.2.1 Coordination with New York State DOT

In order to pursue the project tasks in collaboration with the New York State Department of Transportation (NYSDOT), a meeting was held on December 15, 2011 at Long Island Regional Office (Region 10) in Hauppauge, NY. The purposes of the meeting include:

- Introduce to the agency the weather-sensitive TrEPS model that allows incorporating weather effects in traffic modeling
- Demonstrate how combining weather forecasts with traffic prediction can help the local agency in weather-related traffic management
- Identify the agency's interest for applying these tool (e.g., on-line implementation, off-line experiments)
- Identify a list of WRTM strategies for the analysis and available data

Based on the discussions and suggestions from the meeting, the Northwestern University team prepared a sub-network and selected WRTM strategy scenarios to conduct simulation experiments to assess the effectiveness of different strategies under inclement weather conditions.

6.2.2 Sub-network Preparation

NYSDOT staff and Northwestern University team agreed to extract a smaller network from the entire Long Island area to enhance the estimation and prediction performance of TrEPS during the implementation procedure. The suggested initial sub-network includes the west part of the network, which are bounded by Cross Island Parkway on the west and Sagtikos Parkway on the east as shown in Figure 6-16. Once the rough boundary of the sub-network is determined, the next step is to prepare all the network-related input files based on the new configuration including a new OD matrix which reflect zones and travel demand only for the sub-network area. Figure 6-17 depicts the original Long Island network and the extracted sub-network, and Table 6-5 summarizes characteristics of the two networks.

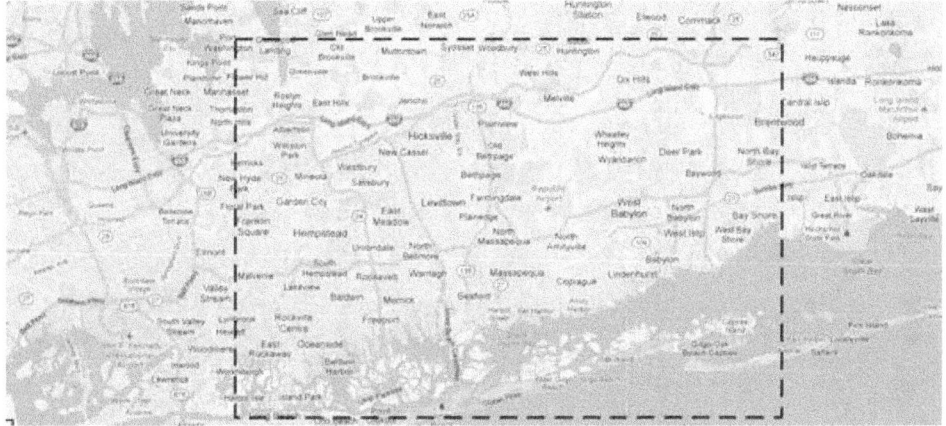

Figure 6-16. Map of the Extracted Network of Long Island (Source: Google Map, Accessed May, 2012)

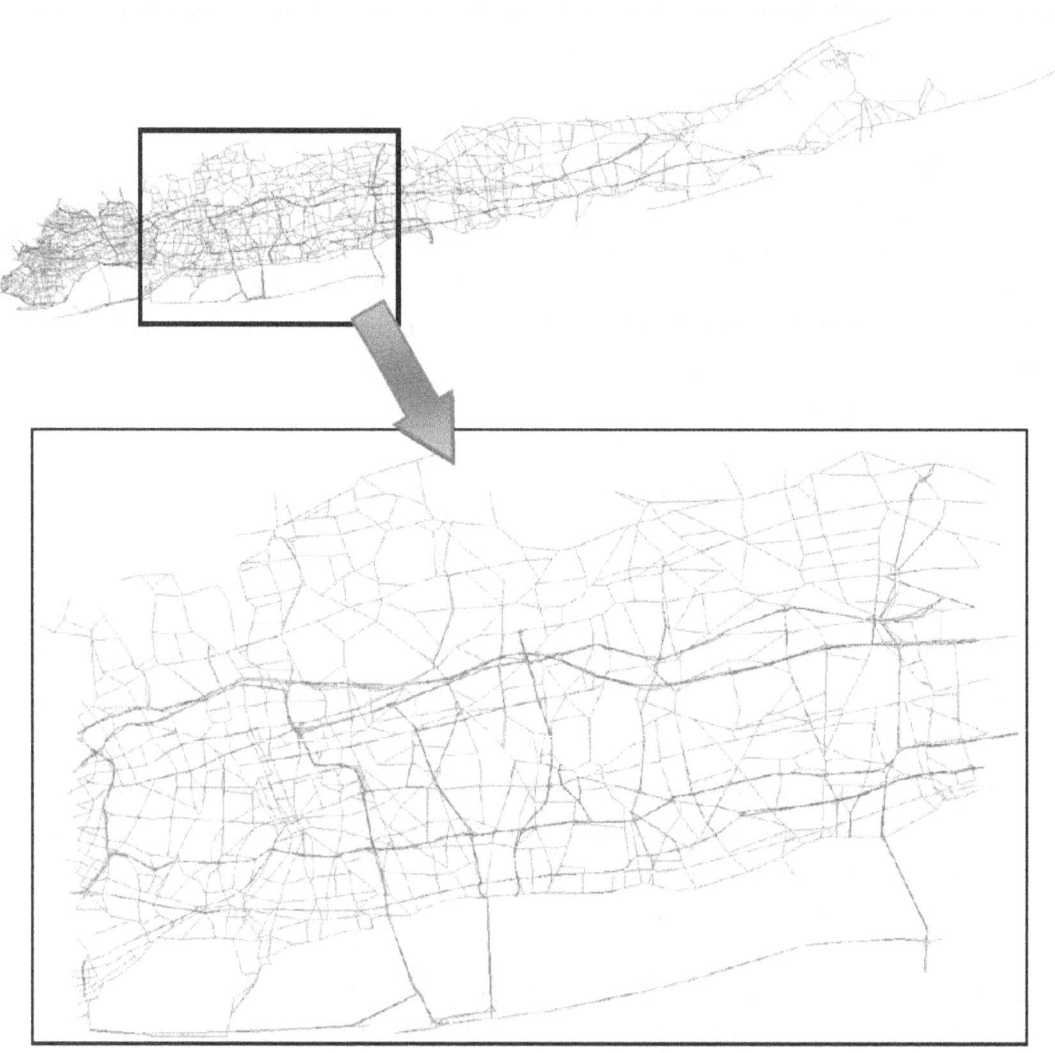

Figure 6-17. Network Extraction for Long Island

Table 6-5. Comparing Network Characteristics for Original and Extracted Networks of Long Island

Network	Original Long Island Network	Long Island Sub-network
Description	<ul><li>1,431 zones</li><li>21,791 links<ul><li>17,945 arterials</li><li>1,588 freeways</li><li>31 highways</li><li>2,087 ramps</li><li>139 HOV facilities</li><li>14 links with tolls</li></ul></li><li>9,402 nodes<ul><li>1,722 signalized intersections</li></ul></li><li>Demand period,<ul><li>6 am– 10 am,</li><li>106 links with observations used in calibration.</li></ul></li></ul>	<ul><li>393 zones</li><li>8,124 links<ul><li>6,153 arterials</li><li>823 freeways</li><li>18 highways</li><li>1,062 ramps</li><li>68 HOV facilities</li></ul></li><li>3,692 nodes<ul><li>582 signalized intersections</li></ul></li><li>Demand horizon<ul><li>6am – 11am</li><li>~783,500 single-occupancy (SOV) passenger car trips,</li><li>~295,500 high-occupancy (HOV) passenger car trips.</li></ul></li></ul>

6.2.2.1 Procedures for Network Extraction and OD Estimation

The network extraction consists of the following three major steps:

Step 3. A new subnetwork is defined by the sets of nodes N, links A and zones Z according to the new boundaries decided by the planners. This subnetwork is designated as the internalnetwork, whereas the remaing sections of the network is designated as the externalnetwork.

Step 4. The original (external+internal) network is simulated in a dynamic simulation and assignment platform (e.g. DYNASMART-P) to obtain the following flows:
   a. External to internal,
   b. Internal to external,
   c. Internal to internal,
   d. External to external (using the internal network),
   e. External to external (not using the internal network).

The sum of these 5 flows defines the time-dependent origin-destination (TDOD) matrix of the original network, whereas the sum of the first 4 flows defines the TDOD matrix of the subnetworkof interest. The outcome of this step is the TDODmatrix of the subnetwork.

These steps finaliz the extraction of the subnetwork with the sets of nodes N, links A, zones Z and the TDOD matrix. Figure 6-18and Figure 6-19 show the time-dependent profiles of extracted travel demand for the sub-newtork for Single Occupancy Vehicles (SOV) and High Occupancy Vehicles (HOV), respectively.

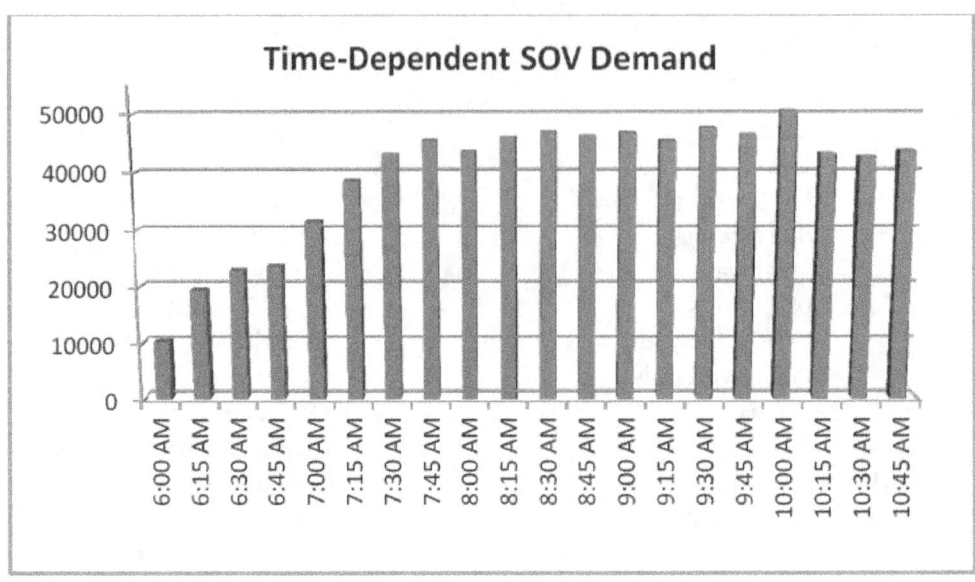

Figure 6-18. Temporal Distribution of 5hr-Demand for Sub-network (Single Occupancy Vehicles)

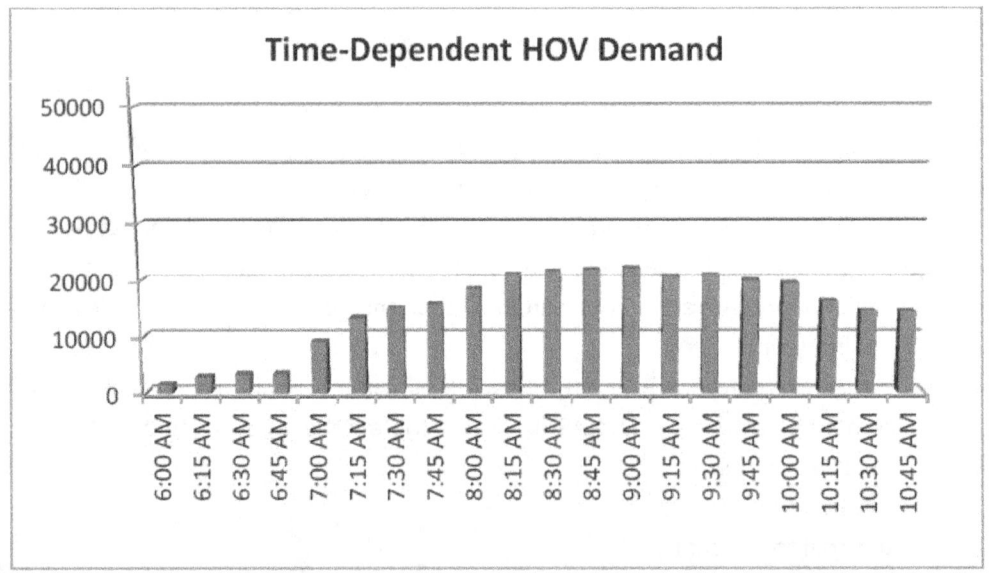

Figure 6-19. Temporal Distribution of 5hr-Demand for Sub-network (High Occupancy Vehicles)

6.2.3 Implementation and Evaluation of WRTM Strategies: (1) Incident Management

6.2.3.1 Weather-responsive Traffic Incident Management

Traffic Incident Management (TIM) is a common operational strategy that is deployed by many agencies to provide the rapid detection, response, and removal of traffic incidents from highways. Many of the strategies that are used to clear incidents from travel lanes can be employed (or expanded) to assist with proactively managing traffic during weather events (*Gopalakrishna et al, 2011*). There are, however, differences between the incident management under normal weather and that under inclement weather conditions. First, as the number of incident is likely to increase under the adverse weather conditions, more service patrols might need to be present in known trouble spots during such weather events. Second, as the response time of recovery vehicles to incident scenes is also affected by the weather condition, locations and service routes of wrecker and pre-positioning patrols would need to be adjusted accordingly.

In addition to developing and expanding the effective incident clearance policies for different weather conditions, weather-responsive traffic management could also involve more proactive incident management strategies, which aim at preventing weather-related incidents before occurrence. By analyzing historical incident data, the likelihood of incident occurrence can be assessed and modeled as a function of the weather condition. Then, whenever adverse weather events occur, traffic management centers might deploy variable speed limit (VSL) and ramp metering strategies to reduce crash risk based on the estimated probability of weather-related incident occurrence.

The development of the above-mentioned weather-responsive incident management strategies can be greatly facilitated through use of the traffic simulation models as various strategies can be tested and evaluated in a simulation environment before adopted in the real-world operations. Once a set of weather-responsive incident management strategies are developed using off-line simulation tools, the TrEPS supports the decision making process for the actual deployment of such strategies in real-time based on the prevailing traffic and weather conditions.

6.2.3.2 Experiment Design

The goal is to demonstrate the use of off-line simulation models in developing and evaluating various WRTM strategies aimed at reducing the impact of weather-related incidents during inclement weather conditions. The selected strategy is Optional Detour VMS available in the DYNASMART simulation model. The Optional Detour VMS represents the variable message signs that display the roadway information (e.g., the traffic congestion ahead) as well as possible detour paths so that drivers could re-evaluate their routes and divert if a better path exists. This optional detour VMS can be used as an incident management strategy, which attempts to prevent serious congestion due to the capacity drop by distributing traffic flow more evenly to alternative routes. The following sub-sections present detailed input scenarios including weather, incident and selected optional detour VMS strategies.

Weather Scenario

For the implementation of the VMS strategy, we construct a snow scenario based on the historical data collected on January 26th 2011 from 6:00 AM to 12:00 PM in Long Island. Figure 6-20depicts temporal profiles of the snow (liquid equivalent) precipitation intensity (inch/hour) and the visibility (mile) of the selected snow scenario. This 6-hour weather scenario is used for the simulation of the sub-network. The entire simulation horizon is 6 hours, in which the first 5 hours are used for generating and moving vehicles based on the estimated OD matrix, which represents the traffic demand between 6:00 AM to 11:00 AM, and the remaining 1 hour is used for simply moving generated vehicles to their destinations.

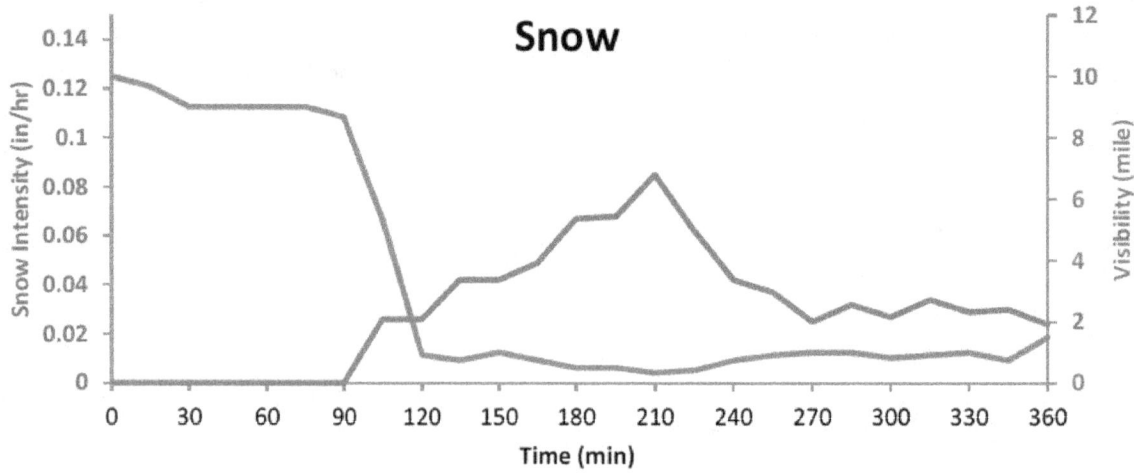

Figure 6-20. Snow Scenario based on Historical Data (extractedfrom2011-01-26 6:00AM – 12:00PM)

Incident Scenario

The incident scenario is constructed based on the actual observations on the snowy day that is selected for the weather scenario. The historical data show that there were three accidents happened along westbound Long Island Expressway (I-495) between 6AM and 12PM on January 26th 2011 as shown in Figure 6-21. Based on the time point at which each accident occurs, we can see that all three accidents occurred during the snow event and, therefore, can be considered as the weather-related incident. Figure 6-22 presents accident locations displayed on the DYNASMART network and

Table 6-6 presents detailed characteristics of each incident. The start and end times represent time points the accidents occur and end, respectively, in the simulation time horizon. The severity represents the fraction of link capacity lost due to the incident. For example, the severity 0.1 indicates that the link capacity decreases by 10% and the remaining capacity becomes 90% of the original capacity.

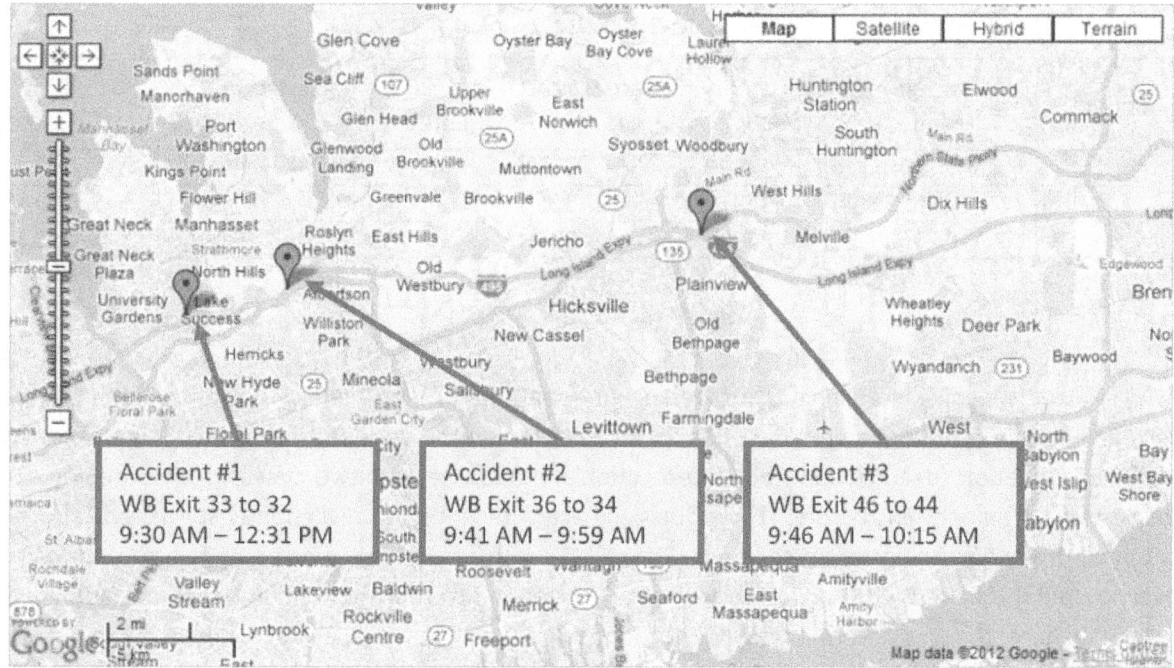

Figure 6-21. Incident Observations on I-495 WB (2011-01-26 06:00AM – 12:00PM) (Source: Google Map, Accessed May, 2012)

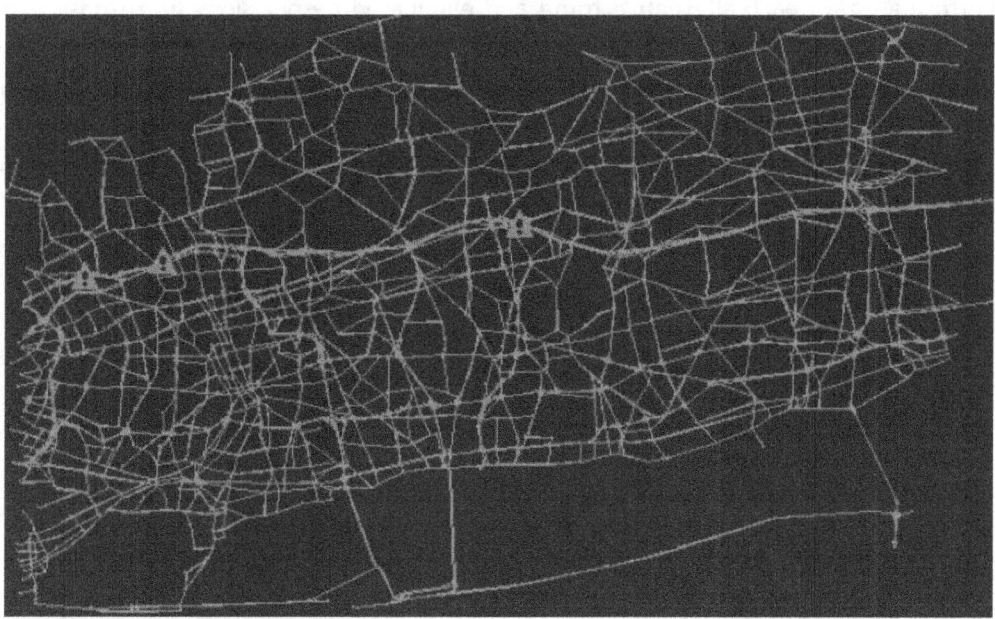

Figure 6-22. Incident Locations displayed in the DYNASMART network

Table 6-6. Incident Scenario based on Historical Data

Accident #	Upstream Node ID	Downstream Node ID	Start Time (minutes)	End Time (minutes)	Duration (minutes)	Severity (capacity loss)
1	22705	22590	210	390	180	0.1
2	23251	32879	220	240	20	0.2
3	25686	25602	225	255	30	0.5

Optional Detour VMS Scenarios

Two variables are considered in constructing the Optional Detour VMS scenarios: the location of detour signs and the VMS response time (i.e., the time gap between the accident occurrence (clearance) and the VMS activation (deactivation). For the location of VMS, we test two types of VMS deployment policies: when an accident occurs, (1) detour signs are activated (i.e., diversion is suggested) at every exit along the adjacent upstream segment (Figure 6-23 (a)); or (2) only selected exits are used for the diversion point (Figure 6-23 (b)). The former case is intended to represent a static type of deployment policies, where the pre-determined VMS locations are used for the strategy deployment. On the contrary, the latter case tries to represent a dynamic type of location selection, where the information on the predicted impact of the strategy on the traffic is incorporated into the VMS location selection. For this, the exits are selected based on the traffic condition on the arterials that will be used for the alternative routes. For instance, after trying the former case (i.e., VMS signs on every exit), we examined the link performance on each alternative route and eliminated detour signs from those exits, whose downstream arterial links already experience a certain level of congestion and do not have sufficient room for absorbing this diverted traffic. For the response time, we also test two different scenarios: (1) the associated VMSs are activated immediately after the accident occurs and deactivated right after the accident is cleared; or (2) the associated VMSs are activated and deactivated 5 minutes after each accident starts and ends.

Based on different combinations of these two variables, four VMS scenarios are constructed as summarized in Table 6-7.

Table 6-7. Optional Detour VMS Scenarios

Scenario	VMS location (exits along the adjacent upstream segment)	VMS Response Time (time delay after the accident starts and ends)
Case 1	All exits	Immediate
Case 2	All exits	5minutes delay
Case 3	Selected exits	Immediate
Case 4	Selected exits	5minutes delay

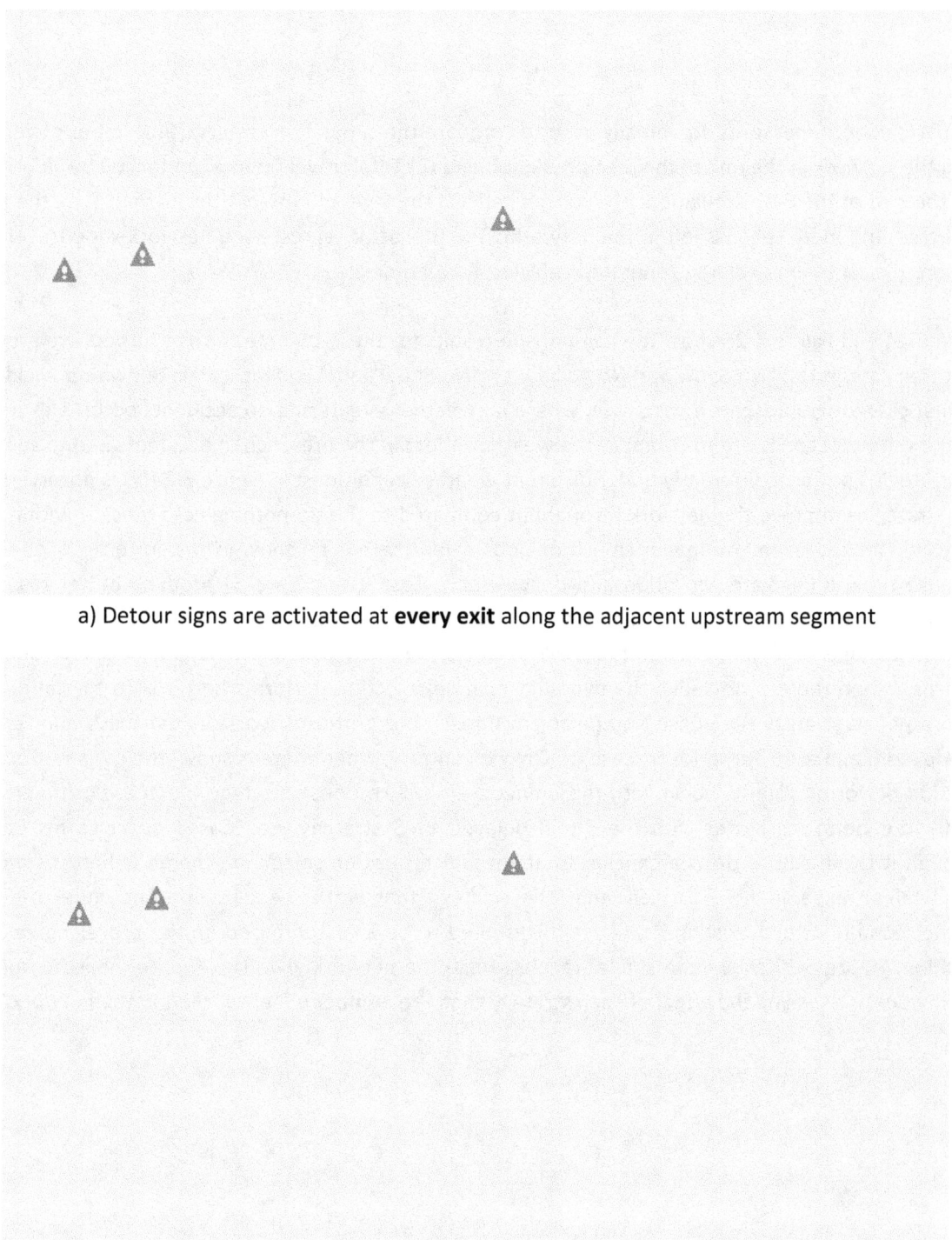

a) Detour signs are activated at **every exit** along the adjacent upstream segment

b) Detour signs are activated at **selected exits** along the adjacent upstream segment

Figure 6-23. Location of Detour Signs for Different Scenarios

6.2.3.3 Analysis Results

Initial Evaluation

Two performance measures are initially used to compare the simulation results: (1) Number of vehicles left in the network at the end of the 6-hr simulation and (2) Total travel times experienced by all vehicles until the end of the 6-hr simulation. The former reflects the level of network throughput (i.e., the more congested, the more vehicles left in the network); and the latter represents a network-wide travel time measure (i.e., the sum of all the generated vehicles' travel times).

Figure 6-24 and Figure 6-25 show the comparison results for these two measures. Outputs are grouped into three categories: "No Snow and No Accident", "No Snow and Accident", and "Snow and Accident." The first category represents a base-case, where no weather events and no accidents occur. The second and the third categories are to compare the weather effect in the presence of accidents and to see how the strategies perform differently under different weather conditions. In Figure 6-24, it is observed that VMS strategies improve the network throughput compared to the do-nothing case (Case0) within each category. The improvement appears much obvious when there is no snow event. In terms of the VMS response time, immediate activation and deactivation (Case 1 and Case 3) produce better results in general, indicating that the prompt incident detection and the immediate response are important in the incident management. In terms of the VMS location, no-snow and snow conditions show different patterns. When there is no snow, the dynamic location selection performs better with the immediate VMS response strategy (i.e., Case3 outperforms Case1), but performs worse when the 5-min delay is introduced (i.e., Case2 outperforms Case 4). On the contrary, when there is snow, the dynamic location selection performs slightly worse with the immediate VMS response strategy (i.e., Case1 outperforms Case3), but performs better with the 5-min delayed VMS strategy (i.e., Case4 outperforms Case2). Although it is difficult to draw a conclusion about which location selection scheme is better from the current small experiment, it is clear that the strategy that works best is different under different weather conditions and existing strategies might need to be adjusted based on the prevailing weather condition. Figure 6-25 shows the similar results, while the network-wide travel time measure appears less obvious in revealing the effect of each strategy than the number of vehicles left in the network.

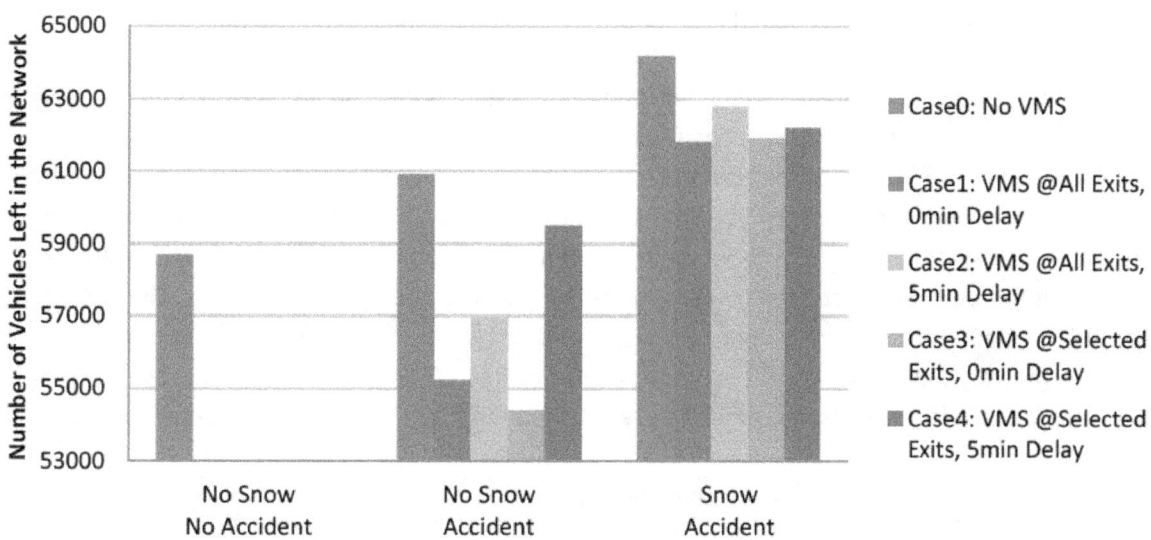

Figure 6-24. Number of Vehicles Remaining in the Network at the End of Simulation

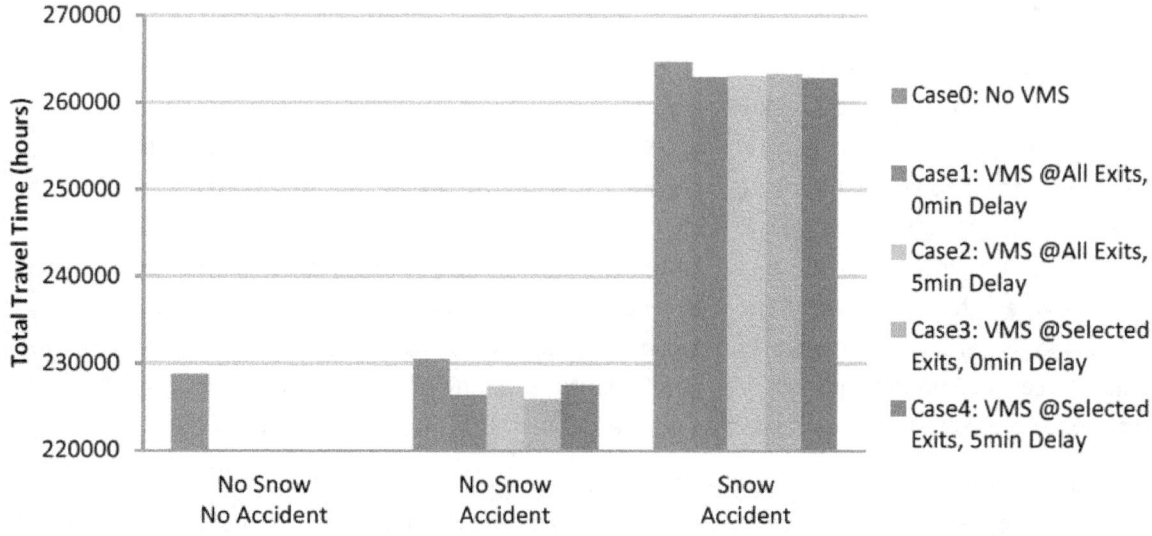

Figure 6-25. Total Travel Times Experienced by Vehicles at the End of Simulation

Detailed Evaluation

Based on the initial evaluation results, four scenarios are selected for a more detailed evaluation. The objective of the detailed evaluation is to better understand the benefits of the WRTM strategies that showed overall good performance in the initial evaluation by performing an in-depth investigation using more diverse performance measrues. The four selected scenarios are presented in Table 6-8, where new scenario names are assigned: "Base" represents the base-case without any weather, incident and WRTM strategy; "SN" includes the snow event, but no accident and no VMS; "SN_ACC" is the scenario with the snow and accident events but without any intervention; and finally "SN_ACC_VMS" represents the scenario with the snow and accident events as well as the VMS intervention this time. The VMS strategy adopted for the fourth scenario is the one that shows good performance improvement under both snow and no snow conditions (i.e., Case 3 in Figure 6-24 and Figure 6-25), where Optional Detour VMSs are activated on the selected exits based on the predicted traffic states immediately after accidents occur.

Table 6-8. Selected Scenarios for Detailed Evaluation

Scenario	Description			
	Snow	Accident	VMS	VMS Strategy
Base	No	No	No	-
SN	Yes	No	No	-
SN_ACC	Yes	Yes	No	-
SN_ACC_VMS	Yes	Yes	Yes	Selected Exits, 0min Delay

(1) Path Travel Time between Two Points

The first measure used for the detailed evaluation is the path travel time between two specific locations.In order to select reference points, we used information from the 511NY traffic information website (http://www.511ny.org/traffic.aspx), where VMS locations and actual messages displayed on each VMS can be identified. There are a number of VMSs that provide travelers with estimated travel times between the current VMS location and specific downstream points as shown in Figure 6-26. Among such reference points, we selected one section for our analysis and Figure 6-27 presents the detailed location. The section stretches from I-495 WB Exit 49N to Route 106/107 and the average travel time ranges from 7 to 8 minutes according to the 511NY traffic information website.

Figure 6-28 presents simulation results for the path travel time for the selected section. The x-axis represents the departure time, which is the time point at which vehicles enter the start-point of the section (i.e., I-495 WB Exit 49N), and the y-axis represents the average travel time experienced by vehicles to travel to the end-point (i.e., Route 106/107).Scenarios with the snow event show increased path travel time starting from minute 100, which corresponds to the time point where the snow event starts as shown in the weather scenario (see Figure 6-20). The accident event that affects this section is the 3rd instance in Table 6-6, which starts at minute 225 and ends at minute 255, and further increases the average travel time as observed in SN_ACC. The VMS strategy (i.e., SN_ACC_VMS) tends to reduce the variation of the average travel time as it decreases the measure at the second peak while slightly increases it at the first peak in the chart. The benefit of the VMS strategy is observed more clearly in the other two representations of these path-based measures.

Figure 6-29shows the average link travel time per unit distance for each link along the selected path, where the x-axis represents the sequence of links and each link is expressed as a pair of the upstream and downstream node IDs. The average link travel time per mile (minutes/mile) is used for measuring the link performance, which is the inverse of speed, and links with higher levels of this measure can be considered as bottleneck of the given path. Focusing on the accident location (i.e., Link 25686→25602), SN_ACC_VMS improves the link performance compared to SN_ACC by reducing the travel time to the level of no accident condition (i.e., SN). In addition, the VMS strategy also resolves the instance of heavy congestion due to the snow and accident events in Link 25405→25295 as SN_ACC_VMS performs better than both SN and SN_ACC.

Figure 6-30 shows a radar chart that compares travel time characteristics of all four scenarios based on the six descriptive measures including mean, median, standard deviation, the 95th percentile, mean of the worst (i.e., longest) 20% of travel times and mean of the best (i.e., shortest) 20% of travel times. The numerical values of the measures are presented in Table 6-9. In general, smaller area in the radar chart indicates better performance (e.g., shorter mean travel time and smaller standard deviation). Base shows the smallest area and SN_ACC shows the largest area. In terms of the individual attribute, the VMS strategy enhances mostly the reliability aspect of the travel time as SN_ACC_VMS decreases the mean of the worst 20% and the 95th percentile of travel times.

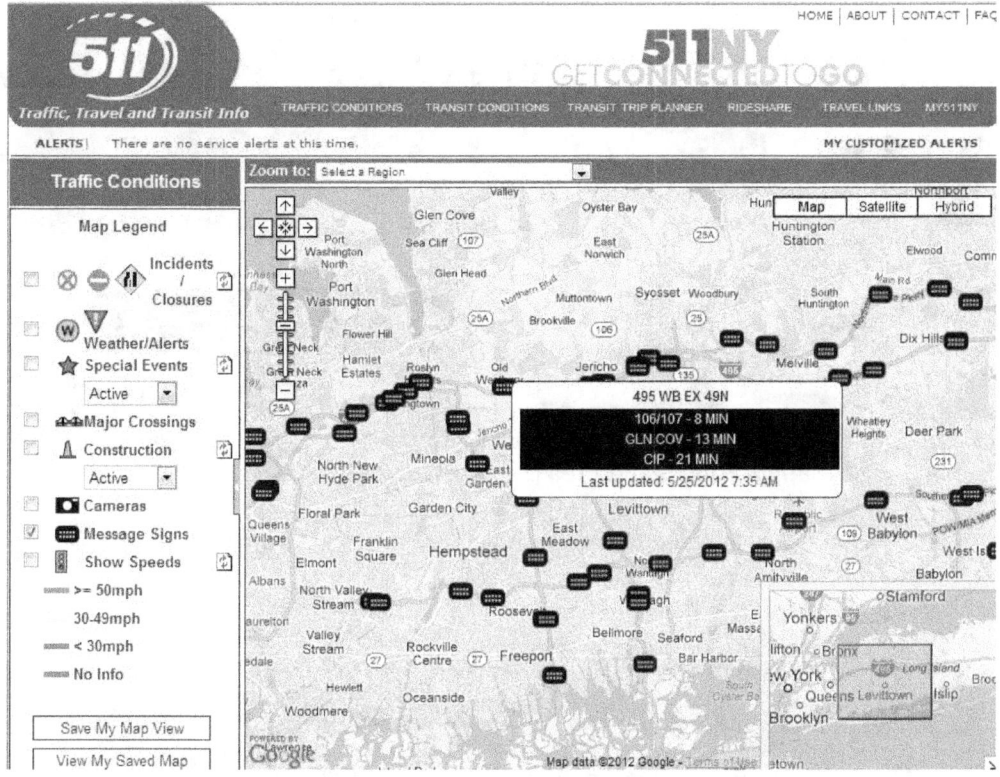

Figure 6-26. 511NY Traffic Information Service (VMS locations and associated messages)
(Source: NY State DOT, http://www.511ny.org/traffic.aspx)

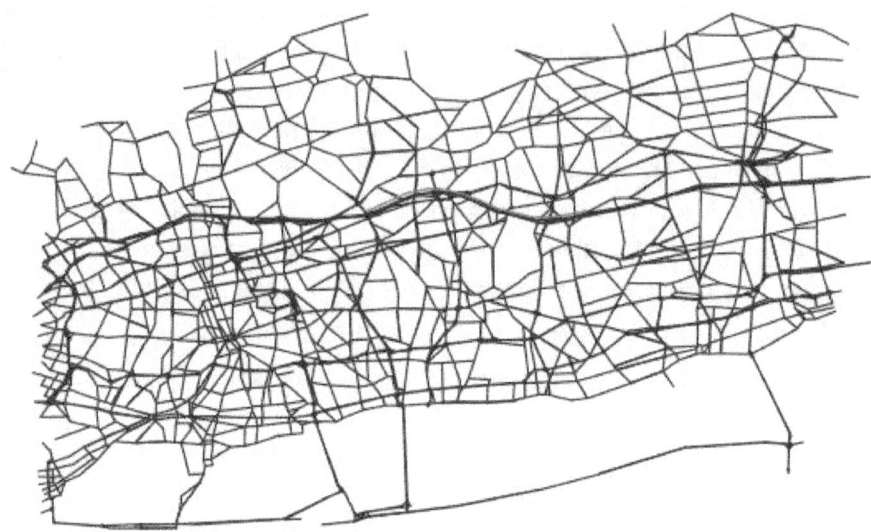

(a) DYNASMART Long Island Sub-network

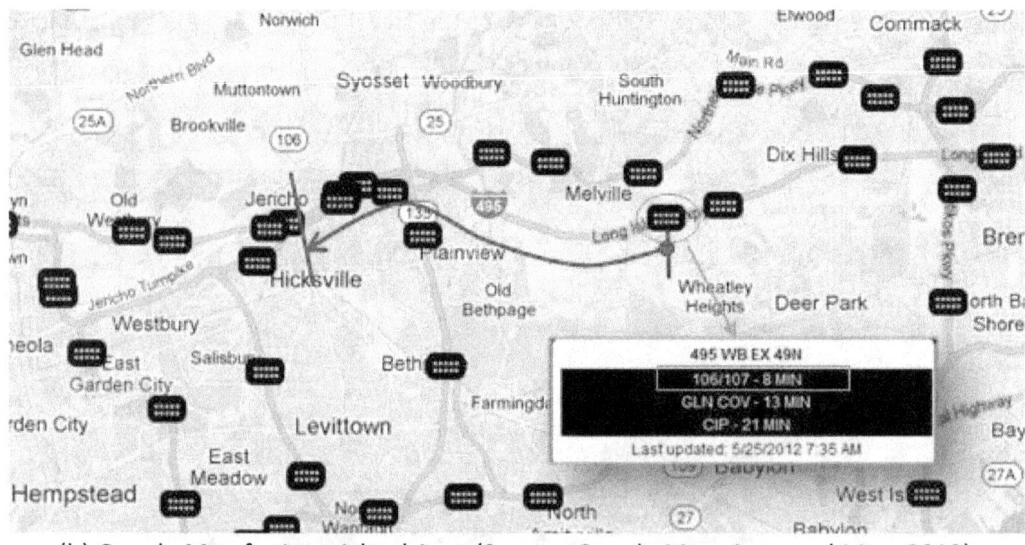

(b) Google Map for Long Island Area (Source: Google Map, Accessed May, 2012)

Figure 6-27. Selected Two Locations for Measuring Path Travel Time

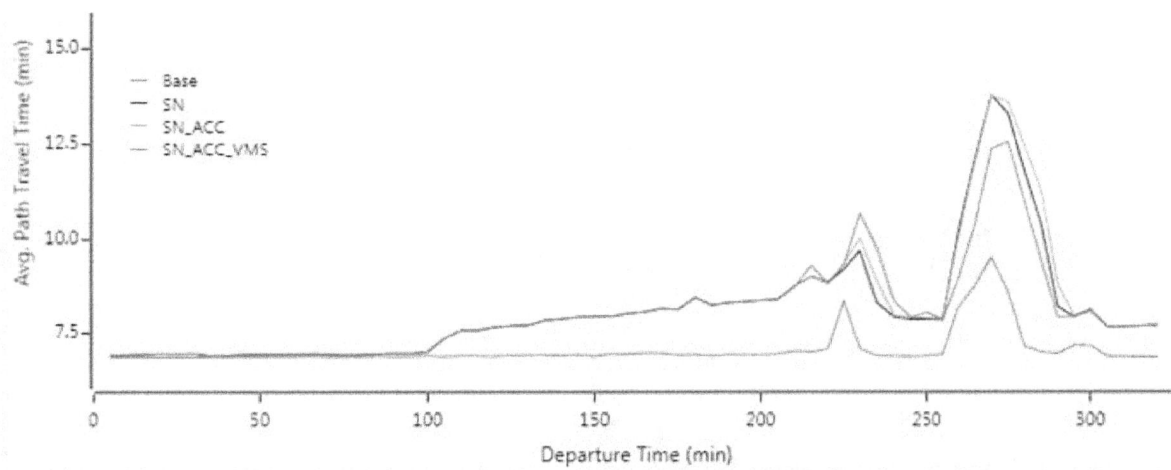

Figure 6-28. Time-dependent Average Travel Time for Selected Path

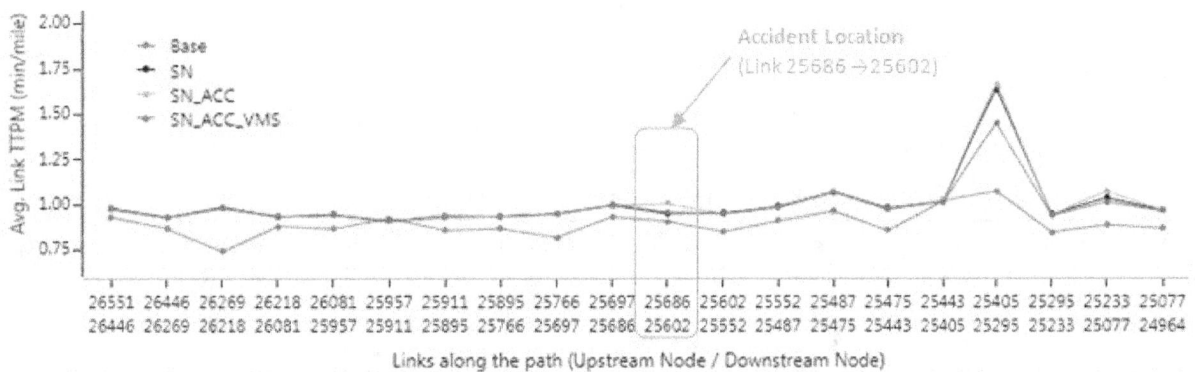

Figure 6-29. Average Link Travel Time Per Mile along Selected Path

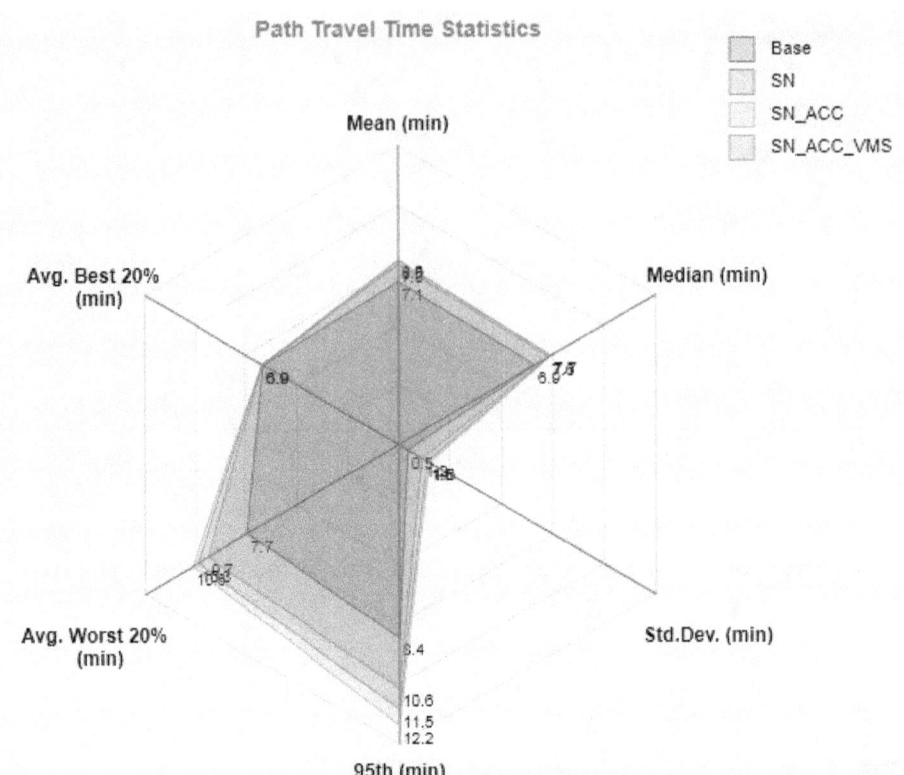

Figure 6-30. Comparison of Travel Time Characteristics of Different Scenarios for Selected Path

Table 6-9. Descriptive Statistics for Path Travel Time

Scenario	#Obs	Mean	Median	Std.Dev	95th	Worst20%	Best20%
Base	3940	7.08	6.9	0.48	8.4	7.71	6.9
SN	3873	7.97	7.7	1.47	11.5	10.28	6.93
SN_ACC	3872	8.03	7.7	1.57	12.2	10.56	6.93
SN_ACC_VMS	3740	7.84	7.6	1.21	10.6	9.74	6.93

(2) Traffic Flows Passing Selected Cross-section

The second measure used for the detailed evaluation is the cross-section throuput, which is measured by the traffic flows passing a selected cross-section.Figure 6-31 shows the selected cross-section on a DYNASMART network and the corresponding Google map for the Long Island area. The vertical bar selects all the westbound links including freeways and arterials. The time-dependent traffic flows aggregated over the selected links are then analyzed. Figure 6-32shows the cumulative vehicle counts and Figure 6-33shows the dynamic flows rate measured by the vehicle counts aggregated over every 5 minutes. The VMS strategy (i.e., SN_ACC_VMS) appears to decrease the vehicle flow rate at first (e.g., around minute 220), but produces the best result in terms of the cumulative number of vehicles at the end (e.g., around minute 360)as it mitigates a sharp drop in the flow rate, which is observed in SN and SN_ACC at around minute 280,and maintains the cross-section throughput at a sustained level.

(a) DYNASMART Long Island Sub-network

(b) Google Map for Long Island Area (Source: Google Map, Accessed May, 2012)

Figure 6-31. Selected Cross-Section for Measuring Traffic Throughput (Westbound)

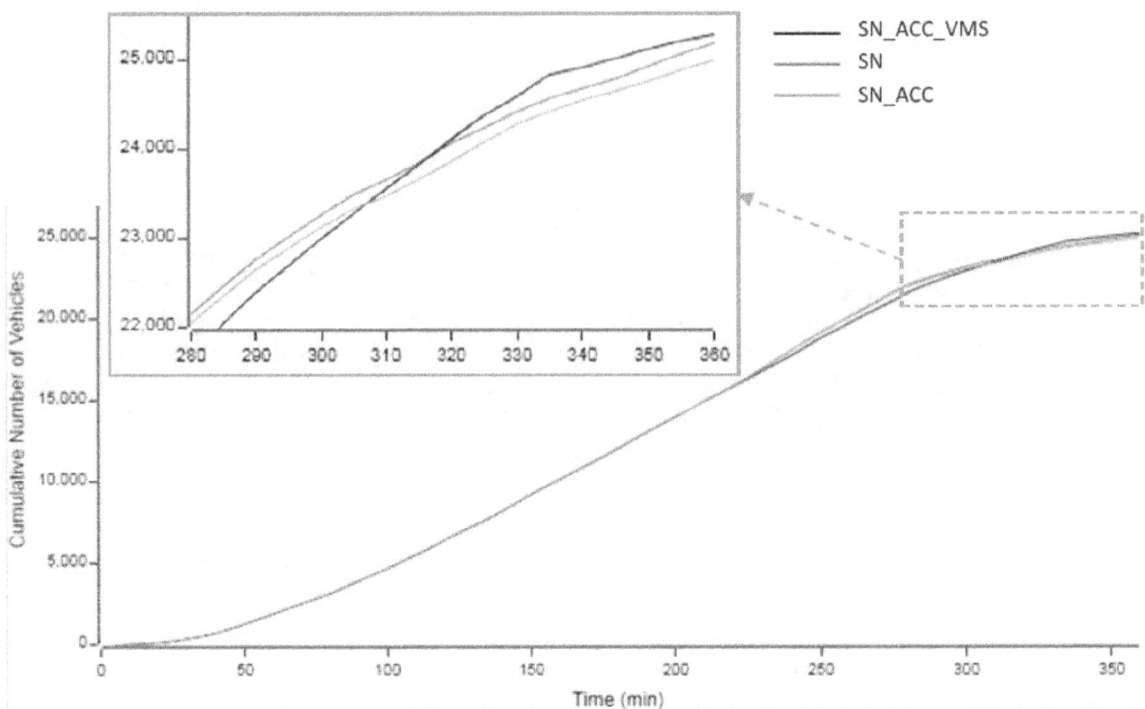

Figure 6-32. Time-dependent Cross-section Throughput Measures (Cumulative Flows)

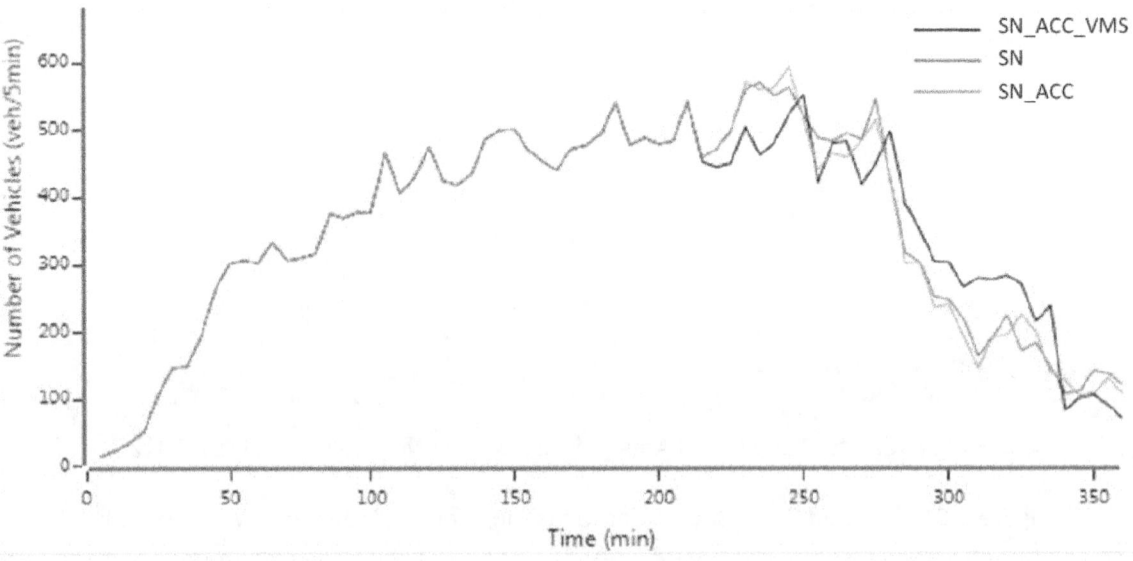

Figure 6-33. Time-dependent Cross-section Throughput Measures (Vehicle Counts/5min)

6.3 Salt Lake City Network

6.3.1 Coordination with Utah DOT

In order to pursue the project tasks in collaboration with the Utah Department of Transportation (UDOT), an initial meeting was held on November 18, 2011 at the Traffic Operations Center (TOC) in Salt Lake City, UT. The purposes of the meeting include:

- Introduce to the agency the weather-sensitive TrEPS model that allows incorporating weather effects in traffic modeling
- Demonstrate how combining weather forecasts with traffic prediction can help the local agency in weather-related traffic management
- Identify the agency's interest for applying these tools (e.g., on-line implementation, off-line experiments)
- Identify a list of WRTM strategies for the analysis and available data

Based on the discussions and suggestions from the meeting, the Northwestern University team has prepared a sub-network and selected WRTM strategy scenarios to conduct simulation experiments to assess the effectiveness of different strategies under inclement weather conditions.

6.3.2 Sub-network Preparation

UDOT staff and Northwestern University team agreed to extract a smaller network from the entire Salt Lake City-Ogden Metropolitan Area to enhance the estimation and prediction performance of TrEPS during the real-time implementation. The suggested initial sub-network includes the Salt Lake City located in the central part of the network from Intersection of I-15 and State Route 89 on the North side and the State Route 145 (Pioneer Crossing) on the South side as shown in Figure 6-34. Table 6-10 summarizes characteristics of the two networks. Once the rough boundary of the sub-network is determined, the next step is to prepare all the network-related input files based on the new configuration including a new OD matrix which reflect zones and travel demand only for the sub-network area.

Figure 6-34. Map of Extracted Network of Salt Lake City

Table 6-10. Comparing Network Characteristics for Original and Extracted Networks of SLC

Network	Original Long Island Network	SLC Sub-network

Description	<ul><li>2,250 zones,</li><li>17,945 links,
	  - 16,291 arterials,
	  - 576 ramps,
	  - 136 highways,
	  - 791 freeways,
	  - 151 HOV facilities,</li><li>8,308 nodes,
	  - 1,023 signalized intersections,</li><li>Demand horizon,
	  - 6am – 9am</li></ul>
		<ul><li>1,284 zones,</li><li>8,292 links,
		  - 7,261 arterials,
		  - 359 ramps,
		  - 12 highways,
		  - 527 freeways,
		  - 133 HOV facilities,</li><li>3,715 nodes,
		  - 442 signalized intersections,</li><li>Demand horizon,
		  - 6am – 10am</li></ul>

6.3.2.1 Procedures for Network Extraction and OD Estimation

The network extraction consists of the following three major steps:

Step 5. A new **subnetwork** is defined by the sets of nodes N, links A and zones Z according to the new boundaries decided by the planners. This **subnetwork** is designated as the **internal** network, whereas the remaing sections of the network is designated as the **external** network.

Step 6. The original (external+internal) network is simulated in a dynamic simulation and assignment platform (e.g. DYNASMART-P) to obtain the following flows:
   a. External to internal,
   b. Internal to external,
   c. Internal to internal,
   d. External to external (using the internal network),
   e. External to external (not using the internal network).

The sum of these 5 flows defines the time-dependent origin-destination (TDOD) matrix of the original network, whereas the sum of the first 4 flows defines the TDOD matrix of the **subnetwork** of interest. The outcome of this step is the **initial TDOD matrix** of the subnetwork.

Step 7. A bi-level optimization algorithm is run iteratively to calibrate the TDOD matrix of the subnetwork obtained at Step 2:
   a. At the upper level of the algorithm; the squared deviations between the simulated and the observed flows on a set of links with traffic counts, and the squared deviations between the **initial** (obtained at Step 2) and the **new** demand values are minimized.

The simulated link flows are defined as a function of the demand values by the so-called **link-flow proportions.**

b. The subnetwork is simulated with the **new** TDOD matrix obtained at Step 3a and the new **link-flow proportions** are calculated for the next run of Step 3a.

Steps **3a** and **3b** are repeated until the squared deviations between the simulated and observed flows converge to a reduced value. This suggests that the extracted subnetwork is capable of reflecting the real-world traffic counts.

This step finalizes the extraction of the subnetwork with the sets of nodes N, links A, zones Z and the TDOD matrix.

6.3.3 Implementation and Evaluation of WRTM Strategies:(1) On-line Implementation

6.3.3.1 Procedures for Connecting Real-time TrEPS to SLC

In order for DYNASMART-X to work in a real-time manner, observations of real-time traffic flow parameters from various type of sensors (including loop-detectors, probe vehicles, etc.) need to be fed into the system continuously and instantaneously. By current design, DYNASMART-X reads real-time data inputs from plain text files with a specific format located on the local machine. A simple external program can serve as bridge between the web-based traffic database (e.g., PeMS) and DYNASMART-X by reading the real-time stream from PeMS and converting it to the specific format that DYNASMART-X recognizes. The procedures to use the real-time traffic data include the following steps:

- Step 1: Get full access to PeMS database that keeps updating real-time traffic data.
- Step 2: Set up look-up table so as to map detector ID from PeMS into DYNASMART network.
- Step 3: Read real-time data stream from PeMS database continuously.
- Step 4: Convert real-time information from the stream to the specific format that DYNASMART-X requires.
- Step 5: Provide the information to DYNASMART-X, either in plain text files (experiment period), or incorporate internally into Data-Management Module of DYNASMART-X.

As a result, DYNASMART-X is successfully connected to the local traffic information and fed the real-time traffic data from detectors in SLC to estimate the current traffic state of the network and predict the future network condition. Figure 6-35 shows the screenshot of DYNASMART-X GUI that displays the estimated prevailing traffic state based on the real-time data and the associated Google traffic map in Salt Lake City. Based on 24-hr demand, which is estimated from an off-line calibration procedure and supplied to TrEPS as a base OD matrix (as shown in Figure 6-36), DYNASMART-X can run 24/7 continuously and produce the estimation and prediction output interacting with the real-time roadway information as well as weather information.

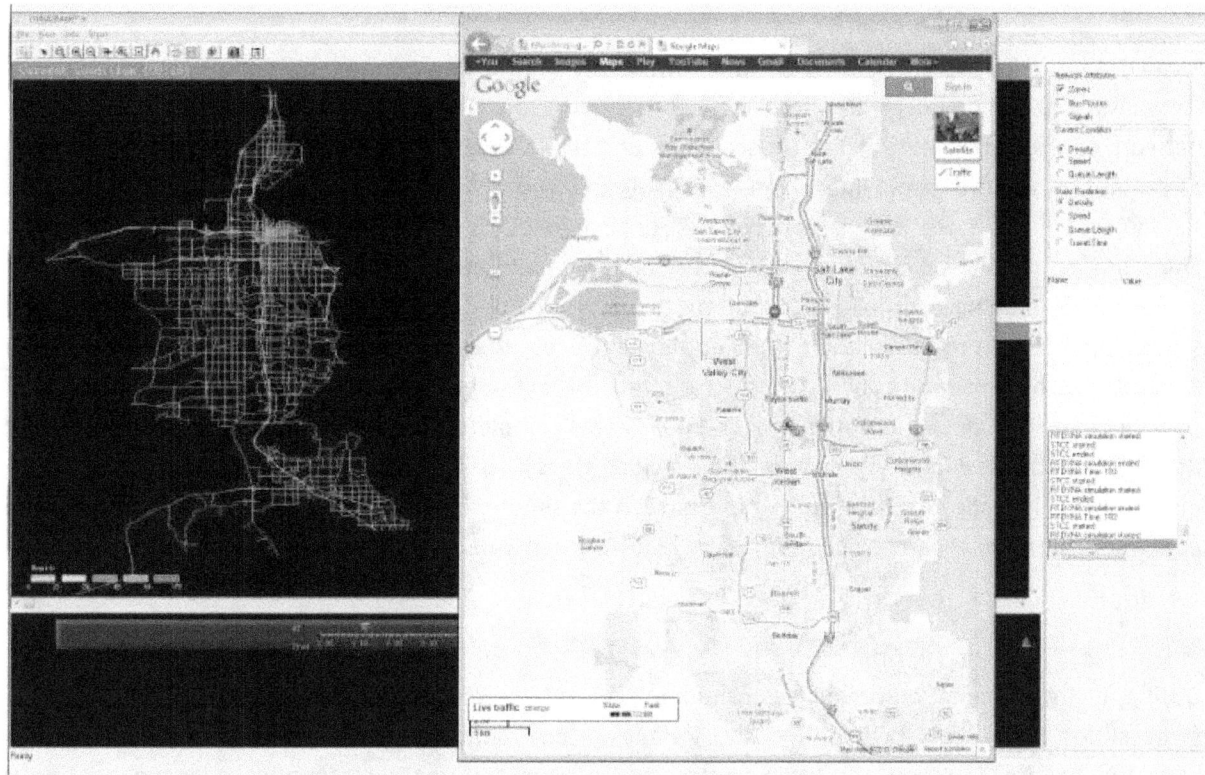

Figure 6-35. DYNASMART-X Connected to SLC Real-time Traffic Data Stream
(Source for top image: Google Map, Accessed May, 2012)

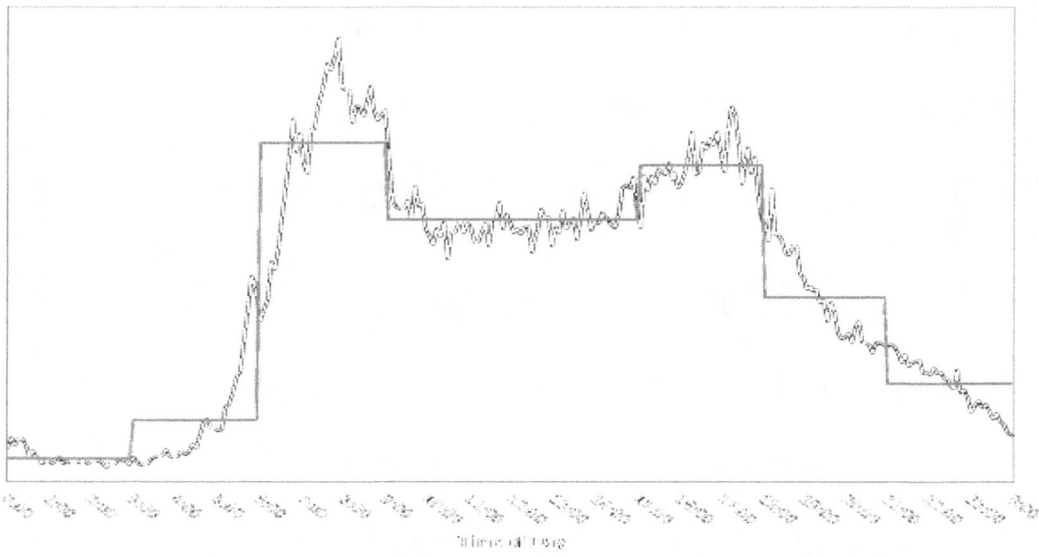

Figure 6-36. Temporal Profile of 24-hr Demand for Salt Lake City Sub-network

6.3.3.2 Procedures for On-line TrEPS Validation

Ideally, all the network components that affect the traffic flow should be supplied as inputs for DYNASMART-X to produces the accurate traffic estimation results. For instance, in addition to the network configuration, external event scenarios and the base OD demand, supply-side operations such as signal timing, VMS deployment and pricing should also be modeled with the corresponding exact schedules to make the simulated roadway environment as close as possible to the real world environment. It is, however, difficult to capture all the details either due to the unavailability of the data or due to the modeling complexity. Based on the data provided by UDOT, we tried to reflect the traffic signal operations in the simulation. This requires a process that maps the traffic signal locations to the DYNASMART network and converts a set of discrete signal timing plans into DYNASMART-specific signal control input files. Once the traffic signal modeling was completed, we validated the on-line TrEPS using the real-time traffic data available from the previous steps discussed above. The validation of the on-line model entails investigating how much gap between the simulation and the observation is generated and how much of the total gap can be reduced by adjusting particular components.

6.3.3.3 Analysis Results for On-line Implementation of TrEPS

During the implementation period (i.e., January, 2012 to May, 2012), the study team monitored weather forecast for Salt Lake City continuously. Whenever precipitation events such as rain or snow are expected on the next day, the team constructed a plausible weather scenario based on a 24-hour forecast of weather conditions and historical observation data to retrieve WRTM strategies for the given weather condition.

As shown in Figure 1-3, the team maintains WRTM Strategy Repository, which is a library that describes available WRTM strategies that might be applied under certain weather conditions. Based on this basic knowledge about which strategies are considered for which weather condition, we retrieved a set of WRTM strategies for the given weather scenario and tested the strategies using the off-line simulation tool, DYNASMART-P, to select one or two intervention strategies that will be used in the on-line implementation.

The results presented in this document are from the procedures conducted in April, 6, 2012 between 6AM and 10AM, where light to moderate snow events occurred in Salt Lake City. The procedures entail the off-line preparation for selecting WRTM strategies and the on-line implementation for testing and evaluating the selected strategies using the real-time traffic information.

Weather Scenario

The weather scenario used for the on-line implementation is presented in Figure 6-37, which depicts the profiles of the snow (liquid equivalent) precipitation (inch/hour) and the visibility (mile). When DYNASMART-X starts at 6AM, an initial weather scenario that is constructed based on the observations up to 6AM and hourly weather forecast for the next several hours is used. As time progresses, the weather scenario is regularly updated using the latest observations and also the latest weather forecast information. The scenario presented in Figure 6-37 is the final weather scenario that is obtained at 10AM.

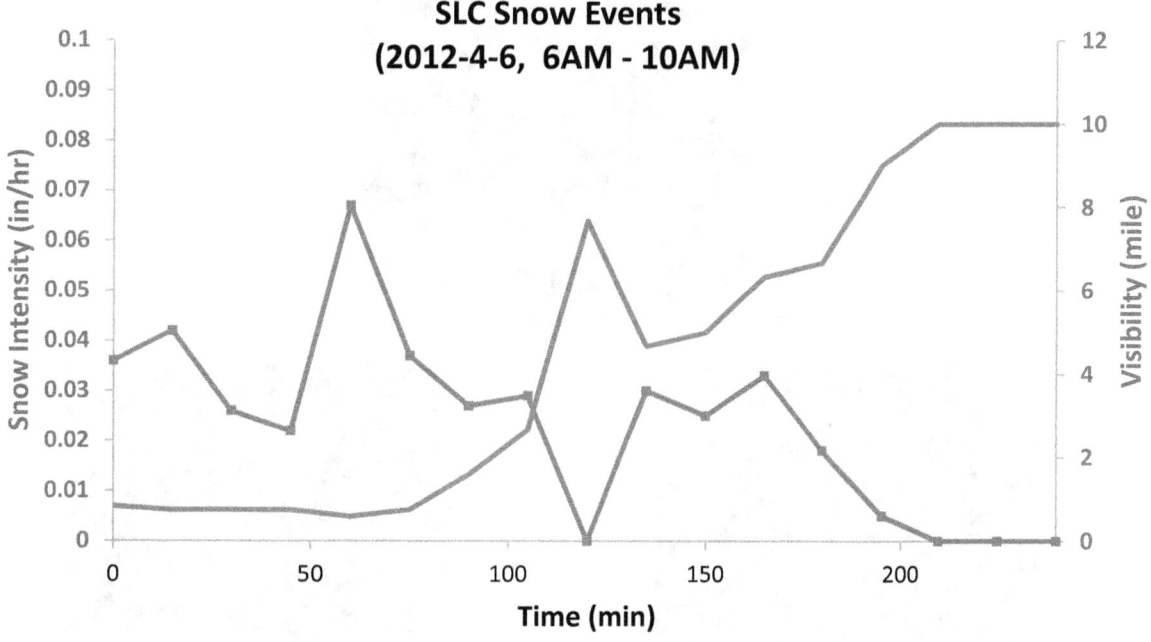

Figure 6-37. Weather Scenario for On-line Implementation in DYNASMART-X

Off-line Preparation (WRTM Strategy Selection)

Once a high probability of precipitation is reported for the next 24 hours, a set of candidate WRTM strategies are prepared in advance (i.e., retrieved from WRTM Strategy Repository using the Scenario Manager). For the snow event on April, 6, 2012, we tested various strategies using the off-line simulation tool and selected Variable Speed Limit (VSL) as the one that is used in DYNASMART-X. Three locations are considered for the VSL deployment as shown in Figure 6-38. Based on the weather scenario prepared for the predicted precipitation event, we obtained the simulation results showing that VSL strategy (Case 2) improves the network performance the most under light to moderate snow conditions as presented in Figure. The figure compares the total travel times produced by five different scenarios including "No Snow, No VSL", "Snow, No VSL" and three VSL strategies; and VSL strategy (Case 2) reduces the total travel time by 35,500 hours compared to the total travel time under "Snow, No VSL" scenario.

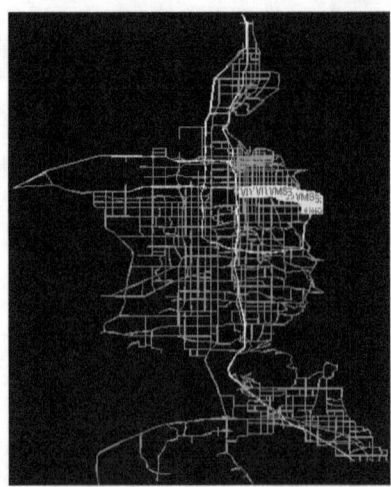

a) Case 1: VSL on Lincoln Highway (Westbound)

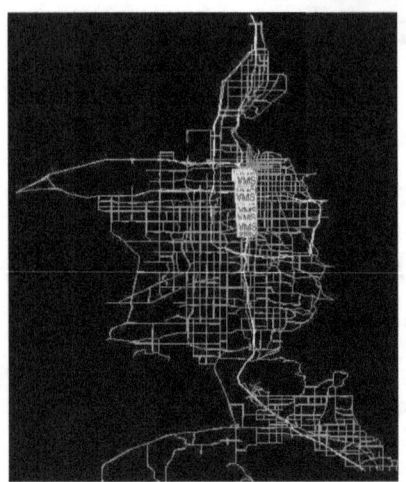

b) Case2: VSL on Veterans Memorial Highway (Southbound)

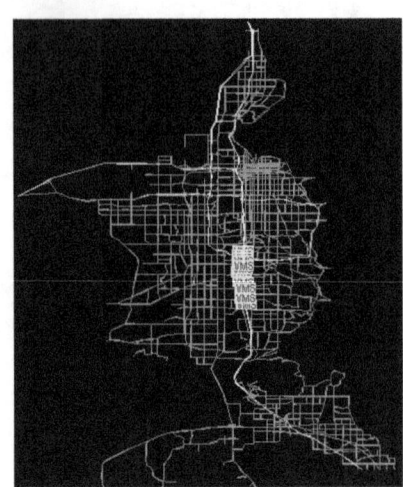

c) Case3: VSL on Veterans Memorial Highway (Northbound)

Figure 6-38. Locations of Variable Speed Limit Signs for Different Scenarios

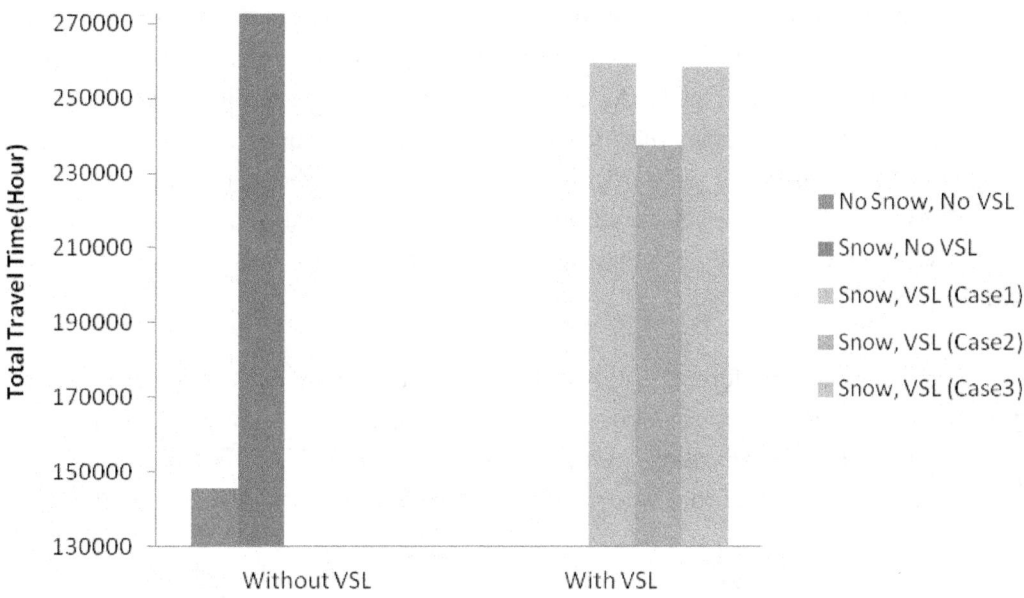

Figure 6-39. Off-line Simulation Results for VSL Strategies

On-line Implementation (WRTM Strategy Evaluation)

We selected VSL strategy (Case 2), where VSL signs are activated on Veterans Memorial Highway (Southbound), as an intervention scenario for the on-line implementation based on the off-line simulation results discussed above.

Starting from 6AM on April 6, 2012, the team executed DYNASMART-X for four hours until 10AM on April 6, 2012. During this period, the real-time traffic stream was fed into DYNASMART-X for adjusting the gap between simulated and observed traffic states. The weather scenario is supplied and updated to reflect the prevailing and forecast weather condition on the network. Based on the estimated traffic state, which is displayed on the RT-DYNA window, DYNASMART-X produces two different future states: P-DYNA0, which is the predicted traffic state under no strategy; and P-DYNA1, which is the predicted traffic state under VSL strategy (see Figure 6-40). At every five minutes, the estimation results are updated in RT-DYNA and new prediction results for the next 20-minutes are presented in P-DYNA0 and P-DYNA1.

Figure 6-40, Figure 6-41 and Figure 6-42 show DYNASMART-X GUI screenshots captured at 7:16AM, 7:48AM and 7:54AM respectively. Each figure presents the network density level using color coding (e.g., red: congested and green: uncongested) and link-specific density for a selected link through a dialog. In the figures, the time point for the current condition and the time points for the state predictions are indicated on the top of RT-DYNA and P-DYNA windows, respectively, in a format of the elapsed time from the start of the simulation. For example, Figure 6-40 shows the current condition for 7:16AM (expressed as 1:16) on the left and the state prediction results for 7:26AM (expressed as 1:26) on the right.

At 7:16AM, we observed that the predicted traffic states for 7:26AM differ between P-DYNA0 and P-DYNA1 (see Figure 6-40).The traffic state under VSL strategy (P-DYNA1) is less congested than that under no strategy (P-DYNA0) as marked as yellow circles in Figure 6-40, where more links are red in the top window than in the bottom window. This indicates that if we had applied the specified VSL strategy at 7:16AM, we would have improved the traffic condition at 7:26AM as shown in P-DYNA1.

Similarly, at 7:48AM, we observed the predicted traffic condition in the same region for 7:58AM (see Figure 6-41) and found that the VSL strategy decreases the number of links with red color compared to the no-strategy case.

As 7:54AM, we examined a different region and observed the predicted traffic condition for 8:04AM (see Figure 6-42). As shown in the yellow circles in the figure, the number of links congested was reduced by the VSL strategy, indicating the positive effect of the VLS strategy after 10 minutes of the deployment (i.e., the traffic state that would have been observed at 8:04AM if we had applied the VSL strategy at 7:54AM).

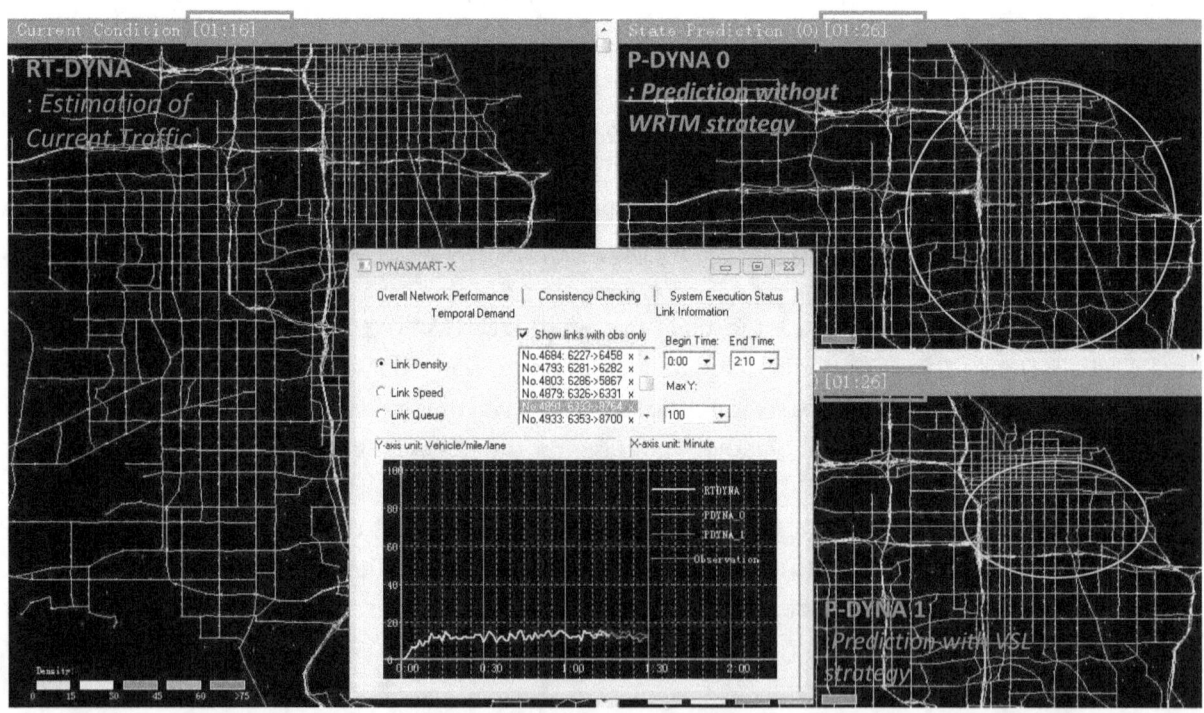

Figure 6-40. DYNASMART GUI during On-line Implementation (at 7:16 AM, April 6, 2012)

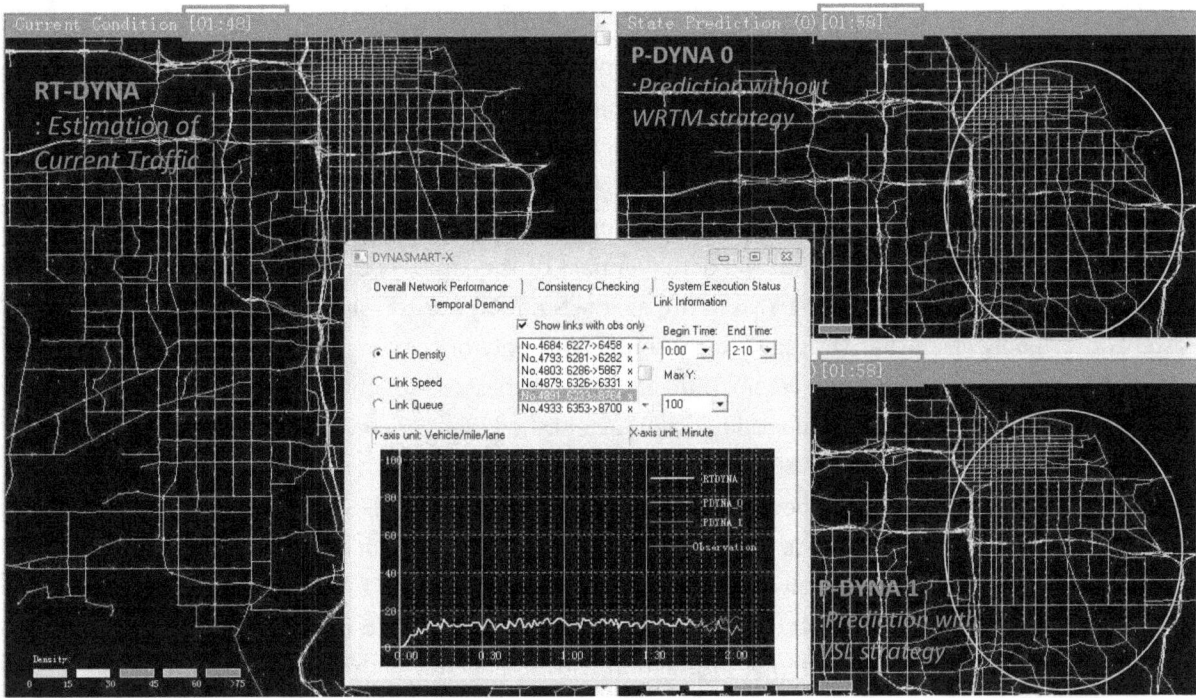

Figure 6-41. DYNASMART GUI during On-line Implementation (at 7:48 AM, April 6, 2012)

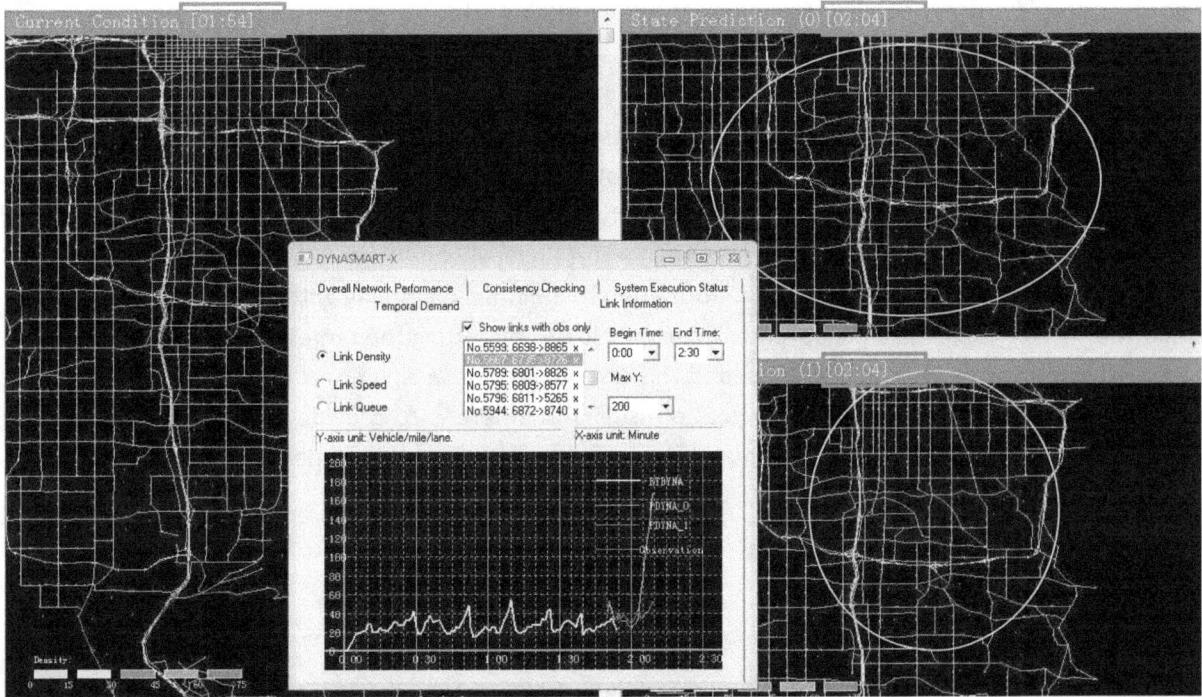

Figure 6-42. DYNASMART GUI during On-line Implementation (at 7:54 AM, April 6, 2012)

Discussion

In this analysis, the team demonstrated the procedures for the on-line implementation and the information that TMC operators can obtain from the real-time TrEPS model, DYNASMART-X, in implementing and evaluating WRTM strategies.

In order to fully evaluate the WRTM strategies using TrEPS, the selected WRTM strategy must be deployed to the real network so that the actual impact of the strategy on the traffic is observed in the detector data; and captured by RT-DYNA in the next estimation horizon thereby updating the prediction results accordingly given that strategy are present in the network. As the TrEPS approach for WRTM is still in its initial stage, and the system that allows agencies to directly deploy the WRTM strategy identified from TrEPS to the real-world is not established yet, however, we could only show the predicted benefit of the selected WRTM strategy at each time point as we demonstrated above. As an intermediate step toward the full-fledged implementation of on-line TrEPS, the approach adopted in this project provides an effective way to introduce such capabilities to agencies— with off-site hosting and maintenance of the TrEPS in the initial period, while further customizing features for local needs This is particularly effective for the interactive TrEPS application demonstrated in the present analysis, where lead time of a few hours or days allows interaction with the agency with professionals who can rapidly turn around the analysis runs. It needs to be tested over a longer period to identify the most effective mechanisms for interaction that would eventually lead to a long-term usage model.

6.3.4 Implementation and Evaluation of WRTM Strategies:(2) Demand Management

6.3.4.1 Weather-responsive Demand Management

Managing demand in this study is about providing travelers with information, aiming at a "shift" of their departure times or trip cancelation so that the total travel demand during the peak periods can be reduced. The key research question here is to study how much demand should be reduced under different weather conditions in order to maintain a certain level of network performance. It is critical for the TMC operators to provide "reliable" information to maintain credibility with roadway users. It is also important to try to minimize the potential economic losses by setting the target demand to its necessary level. Attempting to reduce demand beyond this level might cause significant financial loss to the local business and community. As such, the goal of using TrEPS here is to provide TMC operators with the information on the optimal level of demand that can improve the network performance but not affect negatively the productivity under a given weather condition.

6.3.4.2 Experiment Design

Weather Scenario

As the initial motivation for the demand management is winter storms observed in 2010, which lead to UDOT's keen interest in exploring possible response activities to assist in decision making on future snow incidents, we construct a heavy snow scenario based on the historical data collected on December 29, 2010 in Salt Lake City. Figure 6-43depicts the profiles of the snow (liquid equivalent) precipitation (inch/hour) and the visibility (mile) for the generated snow scenario, which is extracted from the time period between 3:30PM and 8:30PM. The first 4 hours out of this 5-hour weather scenario is used for the simulation of the sub-network. The entire simulation horizon is 4 hours, where vehicles are generated and loaded into the network based on the OD matrix, which represents the traffic demand between 6:00 AM to 10:00 AM.

Demand Scenarios

Total twelve demand scenarios are prepared: one for the benchmark case, which is 100% of the demand under the normal weather condition (i.e., no snow); and the other eleven scenarios with different demand levels under the heavy snow condition. For the generation of the eleven scenarios, we start with the full demand (100%) and reduce the total demand by 5% until the reduction percentage reaches 50%. The purpose of this experiment is to answer the question: "how much reduction in overall demand should we try to achieve under a particular weather scenario in order to maintain the same level of service on our network"?

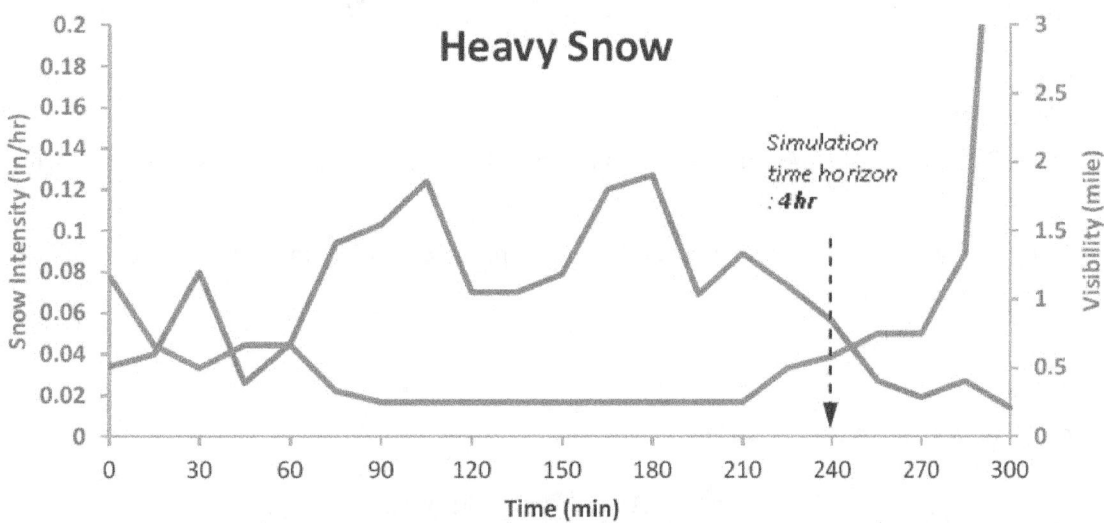

Figure 6-43. Weather Scenario for Demand Management Strategy: Heavy Snow from Historical Data (extracted from 2010-12-29 3:30PM – 8:30PM)

6.3.4.3 Analysis Results

Performance Measures

A set of network performance measures are defined to illustrate network-level traffic conditions under different weather and demand scenarios.

- *Accumulated Percentage of Out-Vehicle*

The accumulated percentage of out-vehicle is the percentage of vehicles arriving at their destinations from the start of the simulation till a given time stamp t. It can be expressed in the following form:

$$\%Accumulated\_Out\_Veh_i^t = \frac{Out\_Veh_i^t}{Tot\_Veh_i^t} \times 100 \qquad (6\text{-}4)$$

Where

$Out\_Veh_i^t$ — Accumulated number of vehicles arriving their destinations from time 0 till time *t* in scenario *i*

$Tot\_Veh_i^t$ — Accumulated total number of vehicles loaded onto the network from time 0 till time *t* in scenario *i*

- Percentage Change in Average Travel Time

$$\%change\_AvgTTime_i^k = \frac{\Delta AvgTTime_i^k}{AvgTTime_{base}} = \frac{AvgTTime_i^k - AvgTTime_{base}}{AvgTTime_{base}} \times 100 \qquad (6\text{-}5)$$

Where

$AvgTTime_{base}$ — Average travel time for full demand in the base case without weather feature

$AvgTTime_i^k$ — Average travel time for *k* percent of full demand in weather scenario *i*

- Percentage Change in Average Stop Time

$$\%change\_AvgSTime_i^k = \frac{\Delta AvgSTime_i^k}{AvgSTime_{base}} = \frac{AvgSTime_i^k - AvgSTime_{base}}{AvgSTime_{base}} \times 100 \qquad (6\text{-}6)$$

Where

$AvgSTime_{base}$ — Average travel time for full demand in the base case without weather feature

$AvgSTime_i^k$ — Average travel time for *k* percent of full demand in weather scenario *i*

Discussion

Figure 6-44 shows the accumulated percentage of out-vehicles representing throughput of the network under different scenarios. One might notice that there are jumps around minute 180 in the chart. This is due to the time-dependent demand and weather patterns (e.g., snow intensity decreases at 180 as shown in Figure 6-43).

Compared to the benchmark case (i.e., Benchmark), where no snow event is present, the snow effect significantly deteriorates the network throughput if the original full demand is used, i.e., no demand management is applied (i.e., Heavy Snow (100% Demand)). It can be seen that the network throughput decreases by about 10% due to weather at minute 180. It is observed, however, the network performance gets better as the demand level decreases. By reducing the demand by 15% (i.e., Heavy Snow (85% Demand)), we can achieve the network throughput similar to the original level (i.e., Benchmark).

Figure 6-45 presents the percentage changes in the average travel time and the average stop time for different demand scenarios relative to the benchmark case. With 20% of demand reduction, both measures are recovered to the level of the benchmark case.

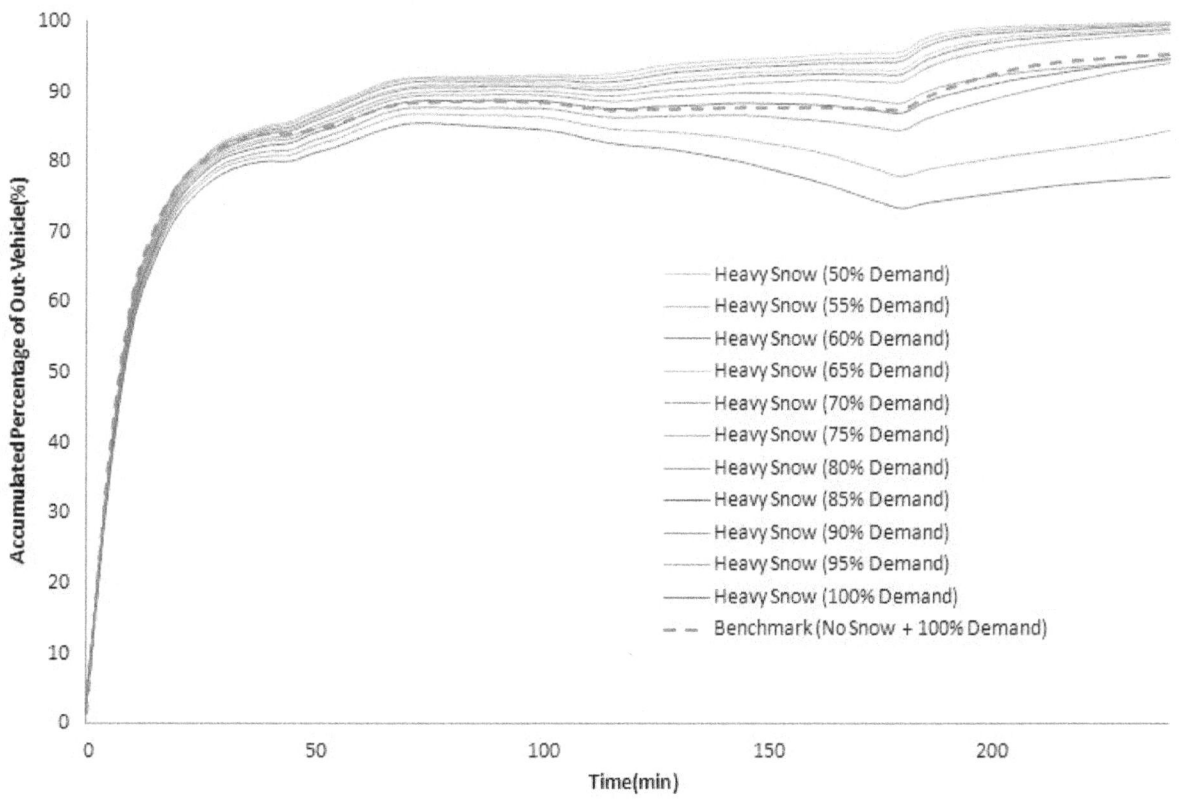

Figure 6-44. Accumulated Percentage of Out-Vehicle for Different Scenarios

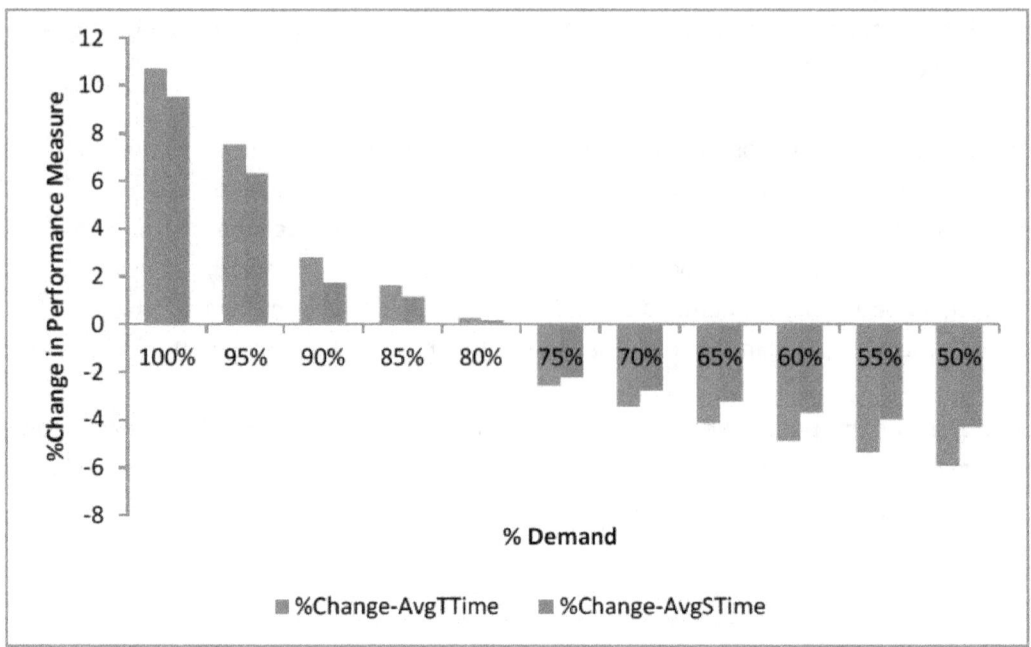

Figure 6-45. Changes in Average Travel Time and Average Stop Time Relative to Benchmark

7. Conclusion

This chapter provides an overall summary of the activities undertaken as part of the study, and highlights the principal accomplishments contributed through the work undertaken in conjunction with this effort. Lessons learned from the project are distilled in the second section; these range from specific findings about the effectiveness of different weather-related traffic management studies, to the process of adoption, use and acceptance of advanced TrEPS methods by the relevant agencies, to effective ways of deploying and using such tools in practice. Building on the study accomplishments and lessons learned, the chapter concludes with recommendations for next steps, both immediate and in the medium term, towards improving the state of the art and practice of WRTM.

7.1 Summary and Accomplishments

7.1.1 Summary

The overall goal of the study is to implement and evaluate weather responsive traffic management strategies using TrEPS models. The TrEPS model selected for this study is DYNASMART-X, which is a real-time system that interacts continuously with loop detectors, roadside sensors and vehicle probes, providing real-time estimates of traffic conditions, network flow patterns and routing information. The high-level framework for incorporating weather impacts in TrEPS developed in a previous project *(Mahmassani et al., 2009)* provided a direction towards the state of the practice in evaluating traffic network performance under adverse weather and developing and deploying weather-responsive traffic control measures. In this project, the methodological development above was further calibrated and validated for locations that experience different weather environments, and embedded in actual procedures used by local and state agencies. To achieve the project's goal, specific methodological approaches were developed and detailed technical activities were undertaken; these are summarized in what follows.

Network Selection

To apply and calibrate the weather-sensitive TrEPS model for major US cities, eight locations were initially considered as candidate sites: Irvine, CA; Portland, OR; Baltimore, MD; New York, NY; Houston, TX; Chicago, IL; Virginia Beach, VA; and Salt Lake City, UT. The team developed a set of criteria for the network selection and examined the characteristics of the candidate cities in various aspects, which include the availability of traffic and weather data, geographic scope, weather pattern, readiness of the simulation network and so on. After conducting a systematic evaluation procedure, the team selected four final study sites: Chicago, IL; Salt Lake City, UT; New York, NY; and Irvine, CA.

Calibration and Validation of Weather-sensitive TrEPS Models
For the selected study networks, traffic and weather data were collected and supply- and demand-side parameters were calibrated. The supply-side parameter calibration includes the estimation of parameters in the traffic flow model (i.e., speed-density relation) and the weather adjustment factors (WAF). The demand-side parameter calibration for this study includes several considerations: the base-case OD matrix estimation, changes in dynamic OD trip patterns due to weather conditions, user responses to information and various advisory/control operations schemes, and so on. In addition, the weather adjustment factors for a selected network (Chicago) were validated by simulating a specific weather scenario with and without using the WAFs. The test result confirmed that the use of WAFs successfully captures the weather effects on both link speeds and flows.

Identification of Existing WRTM Strategies
To identify the existing WRTM strategies, the team conducted a survey of agencies in the four selected study sites. The agencies that provided the responses include Chicago DOT, Utah DOT, Salt Lake City Transportation Division, NY State DOT, NY City DOT, and The City of Irvine. Based on the initial survey results, the agencies and the recommended WRTM strategies to be implemented were identified for three networks, as Irvine does not implement specific weather-responsive traffic management. The agencies that agreed to participate in the study include Chicago DOT, NY State DOT and Utah DOT.

Implementation and Evaluation of WRTM Strategies Using TrEPS
The team established plans for the implementation and evaluation of WRTM strategies for each network in coordination with the associated agency. The plans are categorized into two types: off-line simulation experiments using a weather-sensitive DTA planning tool to test and develop new strategies for future use; and on-line implementation of existing strategies for supporting the decision-making process under inclement weather conditions.

For the off-line experiments, the following strategies were selected for each network:

- Chicago: demand management, variable speed limit, and VMS strategies
- New York (Long Island): weather-responsive incident management using VMS
- Salt Lake City: demand management

For the on-line implementation, the team was able to obtain the necessary resources, particularly the real-time traffic data feeds, for the Salt Lake City network only. This achieved the intended demonstration of the prototype's capability to run with real-time input; however, the absence of significant weather events during the test period, due largely to an unseasonably warm winter, precluded extensive in-situ testing, which remains a topic for future tasks. The team demonstrated a deployment testing model whereby the study team hosted the TrEPS on the NUTC servers, while maintaining communication with the implementing agency. This models provides a template for future implementation-driven development and testing that does not require dedication of specialized staff by resource-constrained implementing agencies.

One of the important study findings is that the primary application of the TrEPS capability lies in the short-term operational planning and preparedness for forecast inclement weather predicted to occur in the next 12 to 48 hours. As such, for each of the networks, the focus shifted on maintaining a calibrated on-call TrEPS model for the extracted subset of the network of interest, as this was the primary interest of the implementing agencies. An essential capability to enable such use is that of the scenario manager prototyped during the course of the present study, as discussed below. However, hooking up the TrEPS to the real-time feeds from sensors remains an interest of the two agencies that could not provide this capability during the limited time frame of the present effort, and as such would be a target of opportunity for future development and testing.

7.1.2 Key Accomplishments of the Study

The study provides an important milestone in the development and application of methodologies to support WRTM. It brings WRTM applications into the mainstream of network modeling and simulation tools, and demonstrates the potential of both WRTM for urban areas and states, as well as of TrEPS tools to evaluate and develop strategies on an ongoing basis, as part of the routine functions of planning and operating agencies.

In addition to achieving and exceeding the intended study objectives, which are to implement and evaluate WRTM strategies using TrEPS models, several contributions and new developments were accomplished as part of the study. These include:

1. Building a library of calibrated traffic flow relations under inclement weather for different areas of the United States (East and West coasts, Midwest, Mid-Atlantic, Mountain region). These cover different geographic regions with different weather patterns and different driver characteristics, as well as different types of networks and facilities. The range of conditions for which these relations were calibrated allows rapid application and prototyping in different areas around the country, by finding the closest match with the calibrated relations.

2. Development of a prototype scenario manager, intended to facilitate application of the TrEPS for different types of weather and other scenarios. The prototype was populated with the historical ASOS and Clarus databases, allowing easy retrieval of any historically occurring weather scenario for a given area of interest. The manager then automatically translates the weather conditions into a corresponding scenario file for the TrEPS. The feedback received during the study from agency personnel suggests that the availability and further development of such a scenario manager could go a long way towards the acceptability, usability and effectiveness of WRTM itself and of the TrEPS within the WRTM process.

3. Development of performance measures or *key performance indicators* (KPI's) to evaluate the effectiveness of particular WRTM strategies in a given network. These allow the model user/agency personnel to compare network performance overall as well as for particular

portions of the network, O-D pairs or user segments, with and without WRTM as well as for different WRTM strategies. This provides an understandable method to quantify and characterize the need for and effectiveness of WRTM, and to communicate these impacts to other personnel, decision makers and ultimately system users as part of demand management strategies. These measures are new to the practice, and convey both the dynamic nature of the information, the network context, as well as the potential impact on users.

4. The concept of "equivalent demand reduction" needed to offset network performance impairment introduced by particular inclement weather conditions, and maintain level of service expected under normal weather conditions. The reduction depends on the nature, intensity, severity and duration of the weather conditions. A methodology was developed and demonstrated for Salt Lake City to solve for the equivalent demand reduction. This information provides a practical target to attain through various information dissemination measures, activity cancellation or rescheduling measures, and possible incentive schemes.

5. Demonstration of a novel deployment model under which TrEPS can be initially introduced to an operating agency and maintained as a remote service hosted by a different organization for as-needed access. In this case, NUTC acted as a host and provider of the service, thereby reducing the demands on the agency staff in the initial period, and providing an effective environment for training by doing.

7.2 Lessons Learned and Next Steps

7.2.1 Lessons Learned

Many important findings were reached through this study regarding the role that network models and simulation methodologies can play in the further development and deployment of WRTM strategies, and the process through which such tools could be most effective in helping agencies attain their objectives within available resources.

1. Most agencies in states and regions that experience severe weather of one type or another believe there is a need for methods to help predict the impact of weather on operations, and develop plans to mitigate the disruptive impact of such weather. All the communication the team had with various agencies around the country found welcoming and receptive attitudes for the tools and their potential for helping improve operations during bad weather.

2. Notwithstanding the perceived need for such tools and interventions, most agencies view WRTM in a holistic manner that entails multiple coordinating activities; traffic management per se is only one of them, and must be integrated with activities such as deployment and scheduling of snow removal equipment, rescue vehicles, and incident management. *This has*

implications for the decision support methodologies developed to support TMC activities. Specifically, it would be *important to consider WRTM along with incident management, achievable through scenario management capabilities,* as this would enhance the effectiveness of both functions. In addition, it is *desirable to incorporate fleet routing and other personnel deployment methods in the same analysis and modeling framework* to allow consideration of more complete weather-related system and operations management scenarios.

3. Needs vary across different agencies and areas depending on factors that include size of the area and demand pressure on the network, and extent to which the population may be used to inclement weather. Similarly, user responses and levels of acceptability and compliance vary accordingly. In our study, we found stark contrasts between Long Island in the New York Metropolitan area and Salt Lake City, UT, in terms of demand adjustment and user behavior in response to agency-supplied information.

4. In all cases, it was evident that the greatest value of the TrEPS methodologies lies in operations planning and preparedness for weather-related events, rather than in minute-to-minute traffic interventions. Given that most weather forecasts can look ahead from a few hours to a few days, with fairly reliable 12 to 24 hours projections, this gives agencies sufficient time to use the TrEPS methodology off-line to predict the impact of the contemplated weather as well as develop the best strategy to mitigate the negative impact. A wider range of strategies are available before the bad weather hits than after it has begun—in particular, demand can be managed more effectively for everyone's benefit. *This has implications for the methodologies and the best manner to utilize them to support WRTM objectives* of the agencies. Specifically, this calls *for an on-call TrEPS that can rapidly retrieve and evaluate a combination of weather-scenarios and WRTM interventions*. Examples of this were provided in connection with the Salt Lake City test deployment. *Additional development and testing of scenario manager features* would go a long way towards attaining the potential of the TrEPS methodologies.

5. An important question and analysis approach that would help most agencies was articulated in connection with the Salt Lake City deployment. Specifically, the SLC agency wanted to know the level of demand reduction that would be necessary under a particular predicted weather scenario to achieve the same level of service attained during normal operations. The study team believes this is one of the most important questions that TrEPS could help answer for agencies concerned with WRTM. This would then form the basis of a multi-modal, comprehensive management approach that would target the quantity and pattern of demand under such conditions. Expanding this type of analysis to more areas, and then testing information strategies to achieve the target level of reduction would provide an important contribution to agencies' abilities to deal with inclement weather.

6. Deployment and maintenance of real-time TrEPS capability requires a commitment of personnel by the agency. The study team has found that the approach adopted in this project provides a

good way to introduce such capabilities to agencies—with off-site hosting and maintenance of the TrEPS in the initial period, while further customizing features for local needs. This is particularly effective for the short-term off-line application described above, where lead time of a few hours or days allows interaction with the agency with professionals who can rapidly turn around the analysis runs. It needs to be tested over a longer period to identify most effective mechanisms for interaction that would eventually lead to a long-term usage model.

7. In addition to the scenario management capabilities introduced in this project and discussed above, the study team has found that it is necessary to devise performance measures that are targeted to the WRTM strategies under consideration. Most agencies have only limited experience with network-related performance measures to start with, and are still on the learning curve in this regard. Grasping and communicating the impacts of the WRTM strategies has called for new performance indicators to be developed, as described in Chapter 5 of the report. These were devised towards the latter stages of the project, as the need for them was identified in the interactions with agency personnel. The study team noticed considerable improvement in ability to communicate and evaluate the impact of the different strategies using these indicators. Accordingly, it would be useful to *further develop and refine these measures, possibly come up with additional performance measures, in conjunction with a test deployment with direct and regular agency interaction and input*. Furthermore, *integrating these measures with a capability for managing scenarios* would contribute considerably to the adoption and effectiveness of WRTM and supporting TrEPS methodology.

8. An important finding that emerged with the expanded set of performance indicators devised for the analysis of the study's evaluation of WRTM effectiveness in different networks is that WRTM not only reduces congestion and average delay to users, but can also considerably improve the *reliability* of travel through the network. This improvement is not only relative to the do-nothing case under inclement weather, but also relative to the situation under normal weather. There is an important story to convey here. Explicit consideration of reliability improvement as one of the consequences of WRTM would help promote its adoption into agencies' existing practices.

9. In all areas, the responses of travelers to information, messages, guidance and controls are an essential ingredient to the overall effectiveness of these measures. These decisions play a central role in the TrEPS methodology, and as such their importance has been recognized early on in the development process. As noted, there is variation in user responses to these measures in different areas. In addition, there are learning effects and other dynamics taking place that will influence these responses—especially as agencies embrace to a greater extent the potential of WRTM strategies to improve operations. While the TrEPS methodology provides the necessary framework and structure to capture these decisions, as well as their evolution, it became clear during the study that a stronger observational basis is needed with regard to what users actually do in bad weather and under different interventions. This need had been

identified in previous studies, though was not directly part of the present scope. The study team believes that a targeted application with tracking of a sample of users would contribute significantly to the ability of agencies to effectively deploy WRTM, and to the tools' usefulness in the analysis and design process for WRTM.

10. In large metropolitan areas, it is important to incorporate alternative transportation modes in the WRTM analysis. In both New York and Chicago, public transit plays an important role in providing and maintaining mobility under most weather conditions. In extreme weather situations, it is equally if not more important to make sure that transit vehicles can move, as it is to ensure that private automobiles can flow. Transit agencies are therefore often partners in clearing certain parts of the road network and effectively managing traffic under such conditions. Similarly, demand management schemes, intended to reduce peak-period demand at times of weather-related traffic capacity reduction, call for information availability on alternative modes. The implication for the TrEPS is to consider transit modes for applications in areas where transit plays a significant part in everyday mobility.

11. The TrEPS analysis methodology developed in this study, and prototyped in the test areas, provides a robust and effective environment for establishing WRTM as an integral element of traffic management in a given area or state. The study demonstrated the flexibility of the methodology, and the considerable scope that remains to build it into a cornerstone and integrated platform for WRTM and, more generally, for system management processes and practices.

7.2.2 Next Steps

In light of the above accomplishments and lessons learned, the study team recommends the following additional steps to build on the findings and accomplishments of the study to further evolve the TrEPS methods into an integrated platform for WRTM, and advance the state of the art and practice of WRTM. These are categorized into immediate steps, which build in a direct manner on the work of the present study, and medium term steps geared towards a more complete methodological capability for WRTM in the context of system management activities.

7.2.2.1 Immediate Steps

1. Implementation-driven development of the scenario manager prototype developed in the present study to support TrEPS deployment and application, primarily in conjunction with WRTM preparedness in response to near-term forecasts of impending inclement weather. This addresses a critical need identified in the course of the present study for having a set of typical scenarios already developed and calibrated in terms of supply and demand characteristics. In particular, O-D demand patterns associated with certain common types of inclement weather could be developed and retrieved according to predicted weather conditions.

2. A test deployment along the remote-hosted model established in the present study, covering a longer term period of three to six months of up time, to include three to five instances of actual inclement weather intervention. This would support the scenario development activities above, and give the testing agency the ability to deploy different strategies as a result of the TrEPS-supported analysis.

3. In conjunction with the above deployment, it would be important to conduct a behavior tracking study that would allow observation of actual user responses to WRTM strategies, with particular focus on demand management strategies. As noted, this is an important gap in existing knowledge, and a critical opportunity from the standpoint of agencies' abilities to mitigate inclement weather. The results would be incorporated in the TrEPS methodology, to improve its ability to predict ways to attain desired demand reduction targets.

4. Develop a template for adoption by operating agencies of WRTM-sensitive TrEPS through an approach that combines the advantages of a remote-hosted calibrated platform, with interactive assistance in the initial stages, towards locally-oriented content for the enhanced scenario manager proposed above. The latter would play the primary role in terms of eliciting and facilitating local engagement in the development and application of WRTM strategies, and in the use of the TrEPS tools for evaluation and decision support. The template would also include a systematic process for monitoring and tracking the value of the TrEPS deployment, particularly through the resulting impact of the TrEPS-enabled WRTM strategies.

7.2.2.2 Medium Term Steps

There exist several opportunities to improve the methodological basis of existing TrEPS methodology, particularly with regard to expanding the range of its usefulness to a more comprehensive scope of WRTM activities. These include:

1. Integration of accident response functionality with WRTM in the real-time TrEPS platform. As noted, while the primary usefulness of TrEPS for WRTM lies in terms of near-term preparedness, within 12 to 48 hours of the onset of predicted inclement weather, the impact of crashes during bad weather is further amplified by the prevailing weather conditions. Accordingly, the online TrEPS would gain in effectiveness if crash responsiveness and WRTM-related functionality are more closely integrated.

2. Similarly, integration of fleet routing functionality, e.g. for snow removal equipment, preventive sanding and freeze-melting agent spreading, and other logistical processes, with the TrEPS platform can greatly enhance the effectiveness of WRTM in the context of overall weather readiness and system management. This would entail incorporating fleet routing and snow-related operations optimization algorithms with the TrEPS-predicted traffic conditions and associated travel times.

3. Along the same lines, incorporating transit-related capabilities is needed to provide essential functionality in larger metropolitan areas with substantial reliance on transit services, or in smaller-sized areas that wish to take advantage of the additional mobility provided by transit during weather-related disruptions.

4. The interplay between off-line scenario development and on-line estimation and prediction, especially with regard to O-D demand estimation and prediction, remains an area where additional development is needed. Previous work has shown that a better starting point, i.e. a priori O-D patterns, can greatly improve the accuracy of on-line TrEPS. This aspect was outside the primary focus of the present study; however, it is an important problem to address, as it bears on the overall accuracy of the TrEPS, and resulting effectiveness of the associated WRTM strategies deployed. This entails revisiting O-D estimation aspects in the on-line TrEPS context, and testing these in connection with an active scenario manager that can retrieve calibrated a priori demand matrices for the particular weather scenarios under consideration.

5. A primary consideration for introducing WRTM, in addition to congestion mitigation, is the concern for motorist safety. Current analysis tools do not consider safety, in the form of crash occurrence or severity, in the context of weather-related scenario analysis. It would be important to enhance the ability of the TrEPS simulation tools to assess the impact on relative safety of inclement weather, and correspondingly the impact of WRTM measures on that important system performance indicator.

6. The potential role of mobile data in connection with TrEPS-based WRTM is under investigation in a separate study. Mobile data holds considerable promise in a real-time setting, though the institutional aspects of obtaining such data remain challenging. Potential providers appear to be willing to make such data available if the financial incentives are there, and if they could be remunerated for the cost of extracting such data and making it available in the desired format. Deployment-based development and testing of the methodology would be improved by incorporating mobile data; however, adequate resources must be provided to enable procurement of such data.

8. References

511NY website, Traffic Travel and Transit Info the New York State Department of Transportation, http://www.511ny.org/traffic.aspx

Alibabai, H., and H. S. Mahmassani. (2008) Dynamic Origin-Destination Demand Estimation Using Turning Movement Counts. *Transportation Research Record: Journal of the Transportation Research Board* 2085, pp. 39-48.

Cambridge Systematics, Inc., *Analytical Procedures for Determining the Impacts of Reliability Mitigation Strategies*, Final Report, SHRP 2 Project L03, 2010.

Cassidy, M. J., and B. Coifman. (1997) Relation among Average Speed, Flow, and Density and Analogous Relation Between Density and Occupancy. *Transportation Research Record* 1591, TRB, National Research Council, Washington, D.C., pp. 1–6.

Cluett, C., D. Gopalakrishna, F. Kitchener, K. Balke, and L. Osborne (2011)*Weather Information Integration in Transportation Management Center (TMC) Operations*, Final Report, FHWA-JPO-11-058, 2011.

Cooper, B.R., and Sawyer, H., (1993). Assessment of M25 automatic fog-warning system (Final report, Project Report 16). Crowthorne: Transport Research Laboratory. TRL-PR-93-16.

FHWA Clarus website, http://www.its.dot.gov/clarus/index.htm

Gopalakrishna, D., C. Cluett, F. Kitchener and K. Balke (2011)*Developments in Weather Responsive Traffic Management Strategies*, Final Report, FHWA-JPO-11-086, 2011

Hogema, J.H. & Horst, R. van der. (1997). Evaluation of A16 Motorway Fog-Signalling System with Respect To Driving Behaviour. *Transportation Research Record* 1573, TRB, National Research Council, Washington, D.C., pp. 63-67.

Hranac, R., et al., *Empirical Studies on Traffic Flow in Inclement Weather*, FHWA-HOP-07-073, 2006.

Ibrahim, A. T., and F. L. Hall. (1994) Effect of Adverse Weather Conditions on Speed-Flow-OccupancyRelationships. *Transportation Research Record* 1457, pp. 184–191.

Jayakrishnan, R., Mahmassani, H.S. and Hu, T-Y. (1994) An Evaluation Tool for Advanced Traffic Information andManagement Systems in Urban Networks. *Transportation Research C*, Vol. 2C, No. 3, pp. 129-147, 1994.

Luoma, J., Rämä, P., Penttinen, M. and Anttila, V. (2000). Effects of Variable Message Signs For Slippery Road Conditions On Reported Driver Behaviour. *Transportation Research Part F*, Vol. 3, pp. 75-84.

Mahmassani, H.S. (1998). Dynamic Traffic Simulation and Assignment: Models, Algorithms, and Applications toATIS/ATMS Evaluation and Operation; in Labbé, Laporte, Tanczos, and Toint (eds.) *Operations Research and DecisionAid Methodologies in Traffic and Transportation Management*, NATO ASI Series, Springer, pp. 104-132.

Mahmassani, H.S. (2001). Dynamic Network Traffic Assignment and Simulation Methodology for Advanced System Management Applications. *Networks and Spatial Economics*, Vol. 1:2 (3), pp. 267-292.

Mahmassani, H.S., Dong, J., Kim, J., Chen, R.B. and Park, B. (2009). *Incorporating Weather Impacts in Traffic Estimation and Prediction Systems*. Final Report, Publication No. FHWA-JPO-09-065; available at http://ntl.bts.gov/lib/31000/31400/31419/14497.htm

Mahmassani, H.S., and Sbayti, H. (2009). *DYNASMART-P Version 1.6 User's Guide*, Northwestern University.

Maze, T., M. Agarwai, and G. Burchett, (2006) Whether Weather Matters to Traffic Demand, Traffic Safety, and Traffic Operations and Flow. *Transportation Research Record: Journal of the Transportation Research Board* 1948 p. 170-176.

Mixon-Hill Inc. et al. (2005) Clarus Weather System Design: Detailed System Requirements Specification. TechnicalReport submitted to FHWA, available at:

http://www.clarusinitiative.org/documents/Final_Clarus_System_Detailed_Requirements.pdf

NOAA National Climatic Data Center (NCDC) website (ftp://ftp.ncdc.noaa.gov/pub/data/asos-fivemin).

Nocedal, J., and S. J. Wright. (2006) *Numerical Optimization*. Springer, 2006.

Nonlinear Estimation: Evaluating the Fit of the Model. Electronic Statistics Textbook. StatSoft, Inc. (2011). Web: http://www.statsoft.com/textbook/nonlinear-estimation/

Pisano, P. and Goodwin, L. (2002). Surface Transportation Weather Applications. Federal Highway Administration(FHWA) in concert with Mitretek Systems, presented at the *2002 Institute of Transportation Engineers Annual Meeting*.

Pisano, P., Alfelor, R.M., Pol, J.S., Goodwin, L.C. and Stern, A.D. (2005). Clarus—The Nationwide SurfaceTransportation Weather Observing and Forecasting System. Presented at the 21st International Conference on Interactive Information Processing Systems (IIPS) for Meteorology, Oceanography, and Hydrology. Available at http://ams.confex.com/ams/pdfpapers/83961.pdf

Rakha, H. A., M. Farzaneh, M. Arafeh, and E. Sterzin. (2008) Inclement Weather Impacts on Freeway Traffic Stream Behavior. *Transportation Research Record: Journal of the Transportation Research Board* 2071, pp. 8–18.

Rämä P. (2001). Effects of weather-controlled message signing on driver behaviour. VTT Publications 447. Espoo: VTT. PhD thesis.

Saberi, M. and Bertini, R.L. (2010) Empirical Analysis of the Effects of Rain on Measured Freeway Traffic Parameters. In Proceedings of the 89th Annual Meeting of the Transportation Research Board, Washington, D.C.

Surface Weather Observations and Reports. *Federal Meteorological Handbook No. 1*. Publication No. FCM-H1-2005, Washington, D.C., 2005.

Verbas, I. O., H. S. Mahmassani, K. Zhang. (2011) Time-Dependent Origin-Destination Demand Estimation: Challenges and Methods for Large-Scale Networks with Multiple Vehicle Classes. Transportation Research Record 4302, pp. 45-56.

Waltz, R. A., and T. D. Plantenga. (2009) KNITRO 6.0 User's Manual. Ziena Optimization Inc.

Zhou, X. (2004) *Dynamic Origin-Destination Demand Estimation and Prediction for Off-Line and On-Line Dynamic Traffic Assignment Operation*. Ph.D. Dissertation. University of Maryland, College Park.

Zhou, X. and Mahmassani, H.S. (2007) A structural state space model for real-time traffic origin–destination demand estimation and prediction in a day-to-day learning framework. *Transportation Research Part B: Methodological* 41(8), pp. 823-840.

Zhou, X., Qin, X. and Mahmassani, H.S. (2002) Dynamic origin-destination demand estimation with multiday link traffic counts for planning applications. *Transportation Research Record: Journal of the Transportation Research Board* 1831, pp. 30–38.

Appendix A: Candidate Networks and Adjacent ASOS Stations

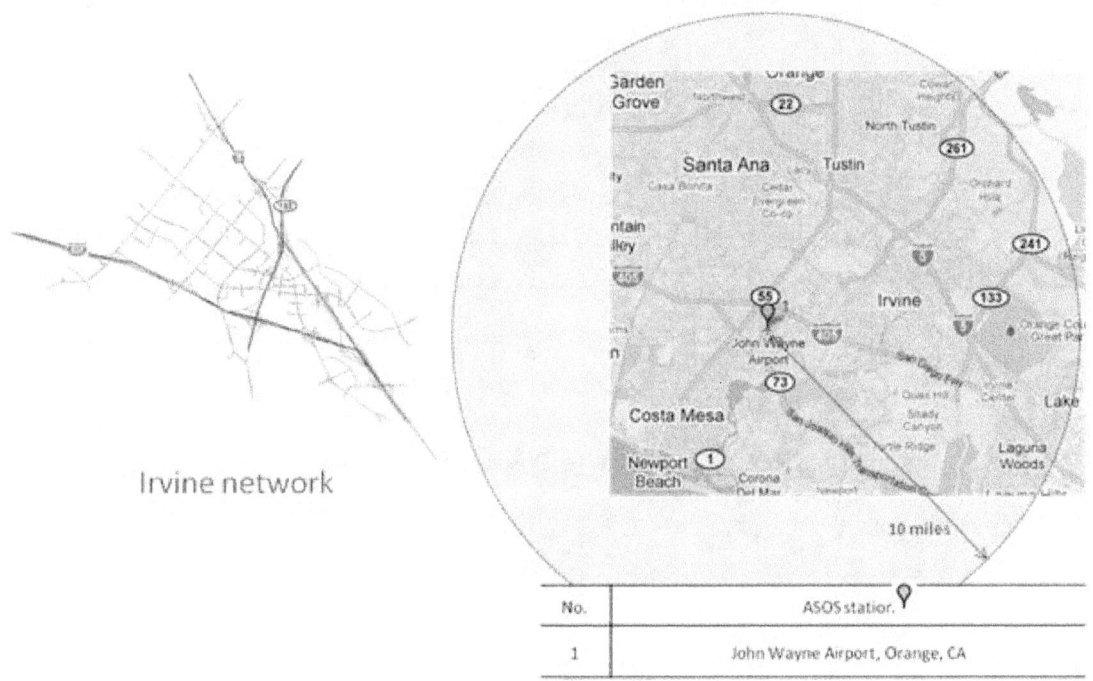

(Source: FAA, Surface Weather Observation Stations)

| Appendix A | Candidate Networks and Adjacent ASOS Stations |

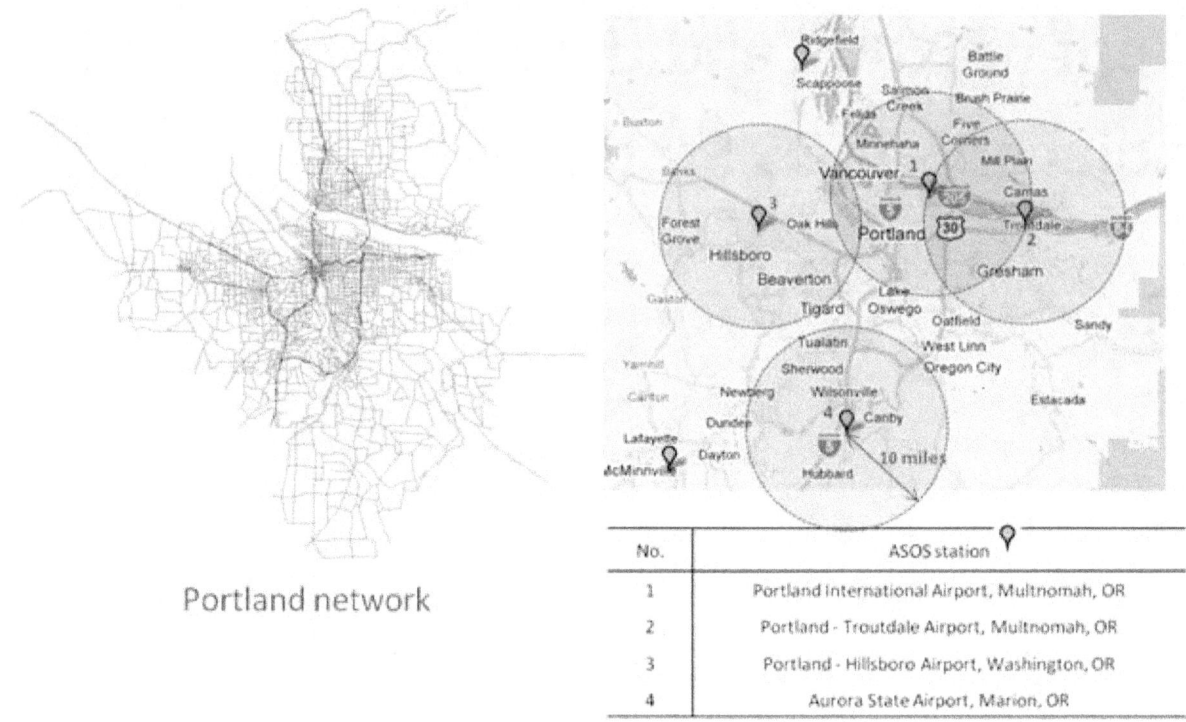

Portland network

(Source: FAA, Surface Weather Observation Stations)

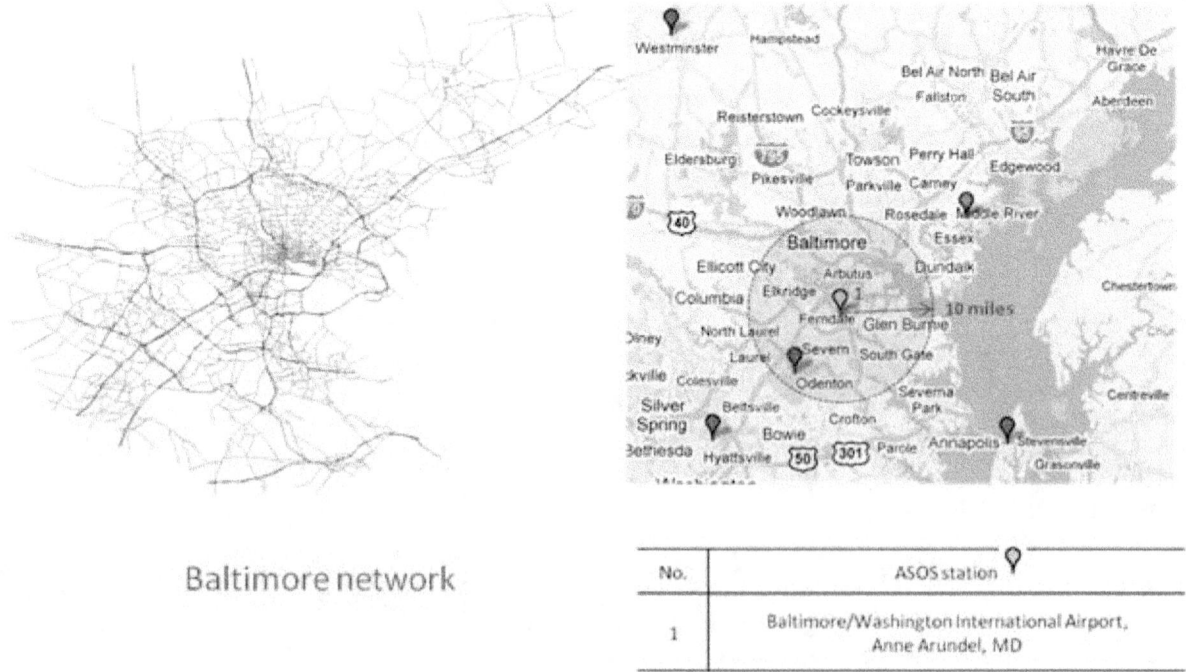

Baltimore network

(Source: FAA, Surface Weather Observation Stations)

| Appendix A | Candidate Networks and Adjacent ASOS Stations |

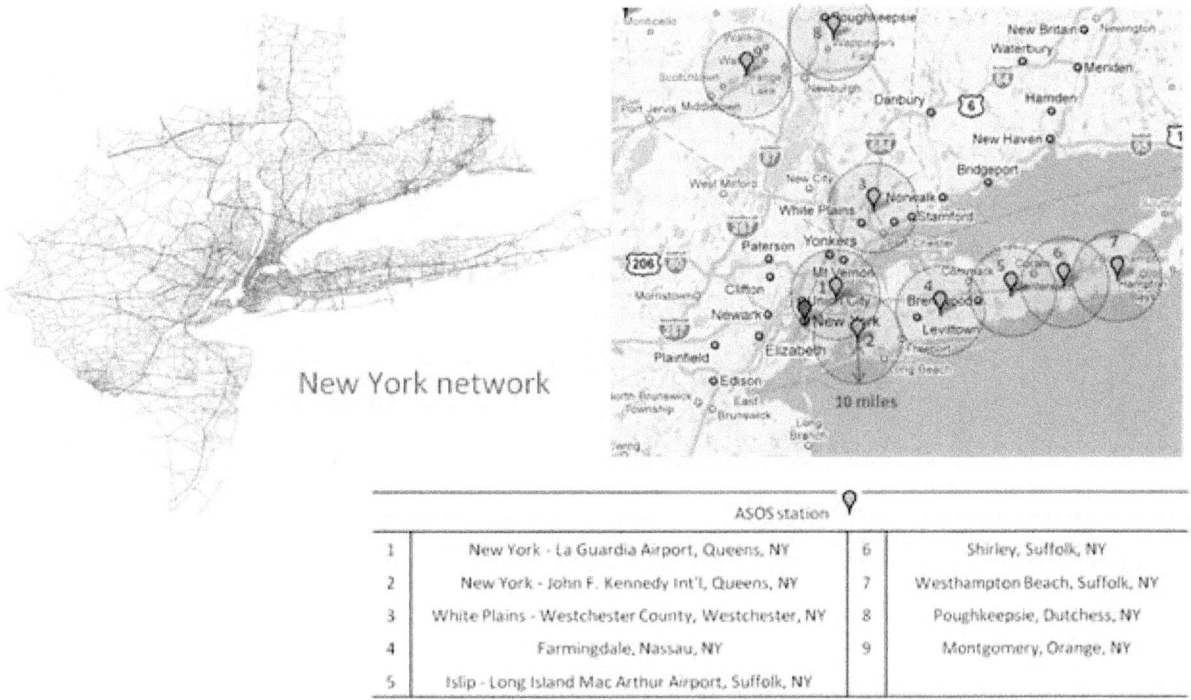

New York network

	ASOS station		
1	New York - La Guardia Airport, Queens, NY	6	Shirley, Suffolk, NY
2	New York - John F. Kennedy Int'l, Queens, NY	7	Westhampton Beach, Suffolk, NY
3	White Plains - Westchester County, Westchester, NY	8	Poughkeepsie, Dutchess, NY
4	Farmingdale, Nassau, NY	9	Montgomery, Orange, NY
5	Islip - Long Island Mac Arthur Airport, Suffolk, NY		

(Source: FAA, Surface Weather Observation Stations)

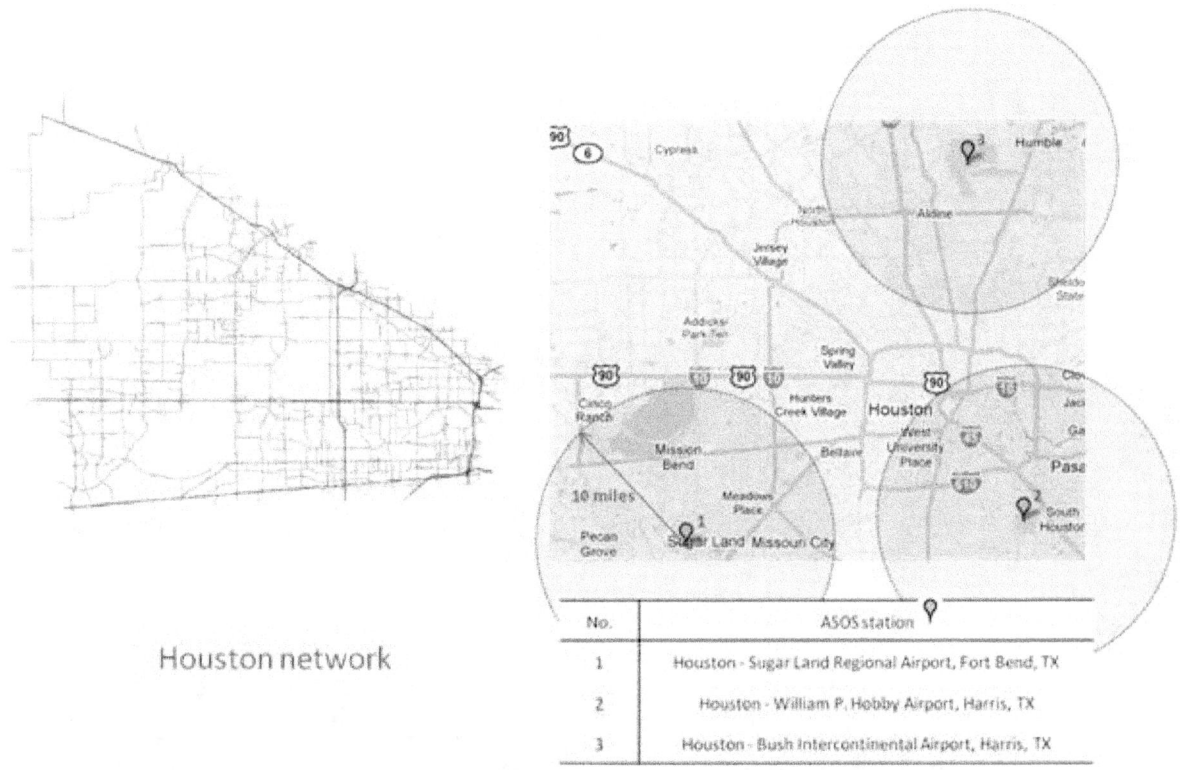

Houston network

No.	ASOS station
1	Houston - Sugar Land Regional Airport, Fort Bend, TX
2	Houston - William P. Hobby Airport, Harris, TX
3	Houston - Bush Intercontinental Airport, Harris, TX

(Source: FAA, Surface Weather Observation Stations)

Appendix A — Candidate Networks and Adjacent ASOS Stations

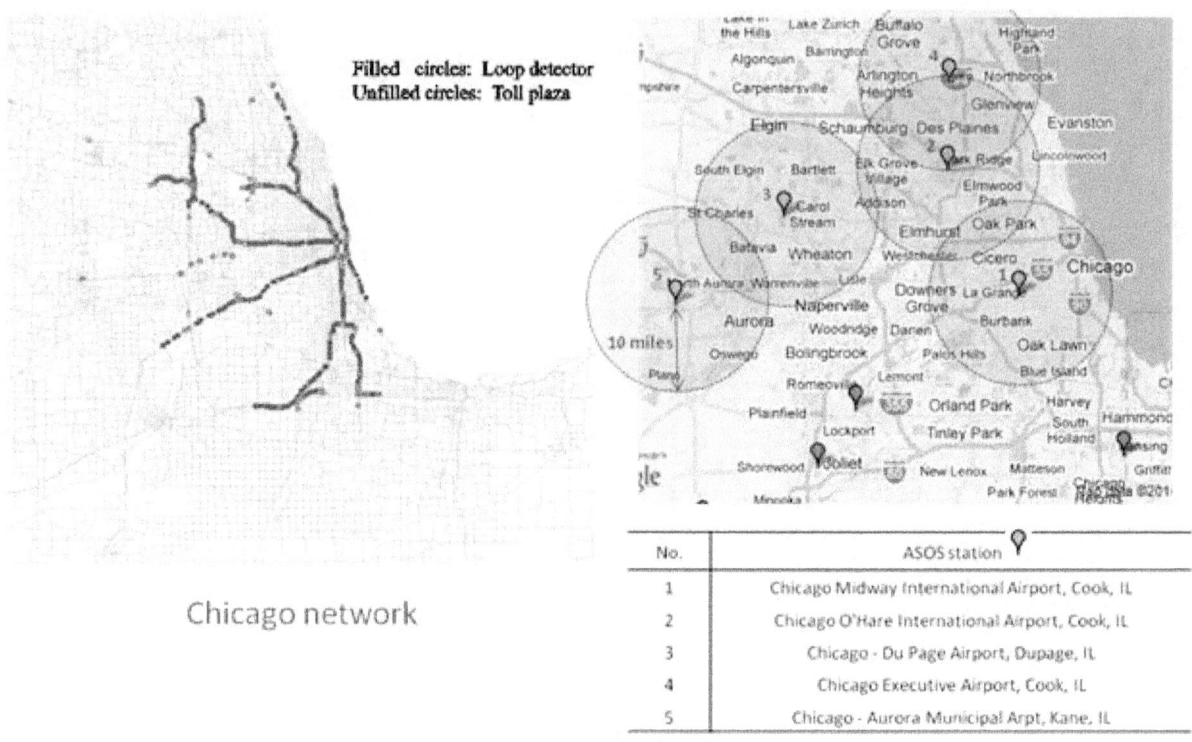

Chicago network

(Source: FAA, Surface Weather Observation Stations)

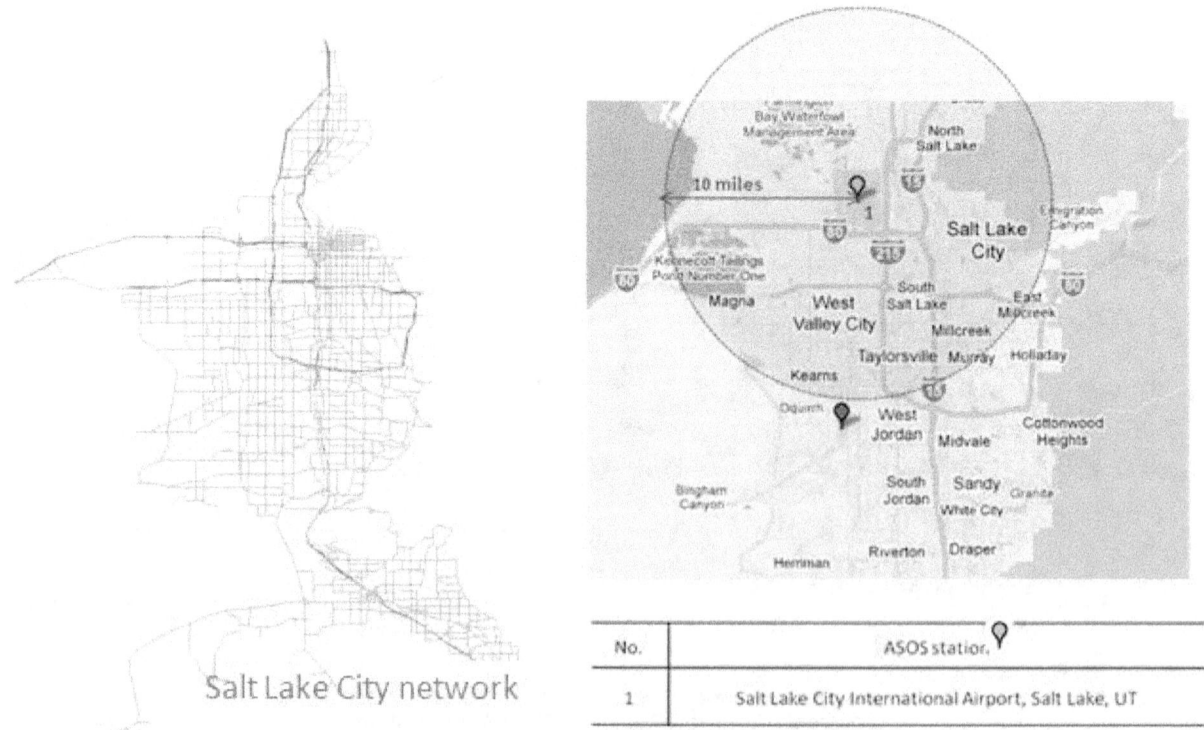

Salt Lake City network

(Source: FAA, Surface Weather Observation Stations)

Appendix A — Candidate Networks and Adjacent ASOS Stations

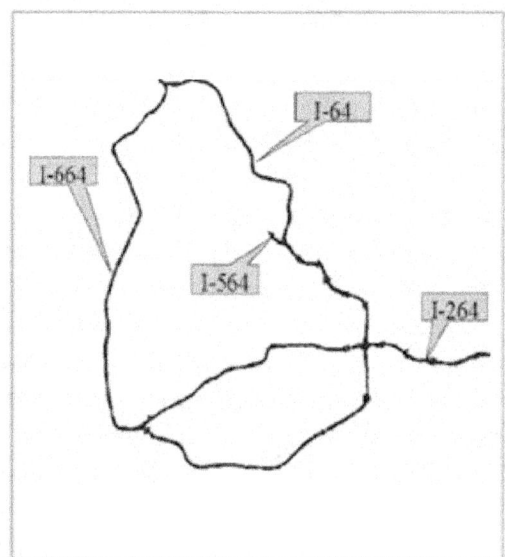

Virginia Beach network

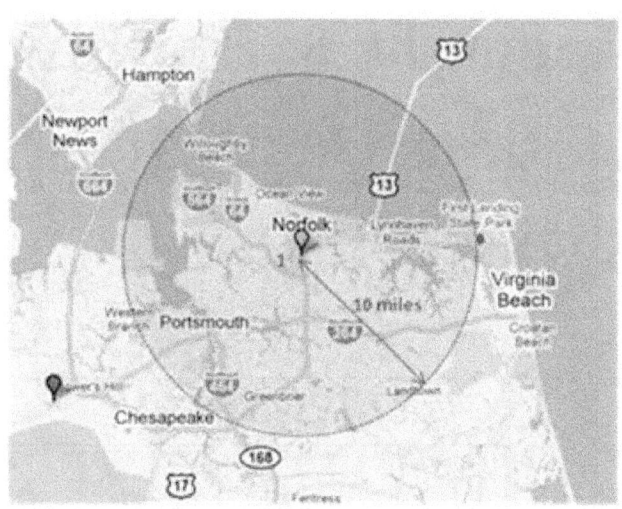

No.	ASOS station
1	Norfolk International Airport, Norfolk City, VA

(Source: FAA, Surface Weather Observation Stations)

Appendix B: Distribution of Weather Stations (ASOS Stations vs. Clarus System(ESS))

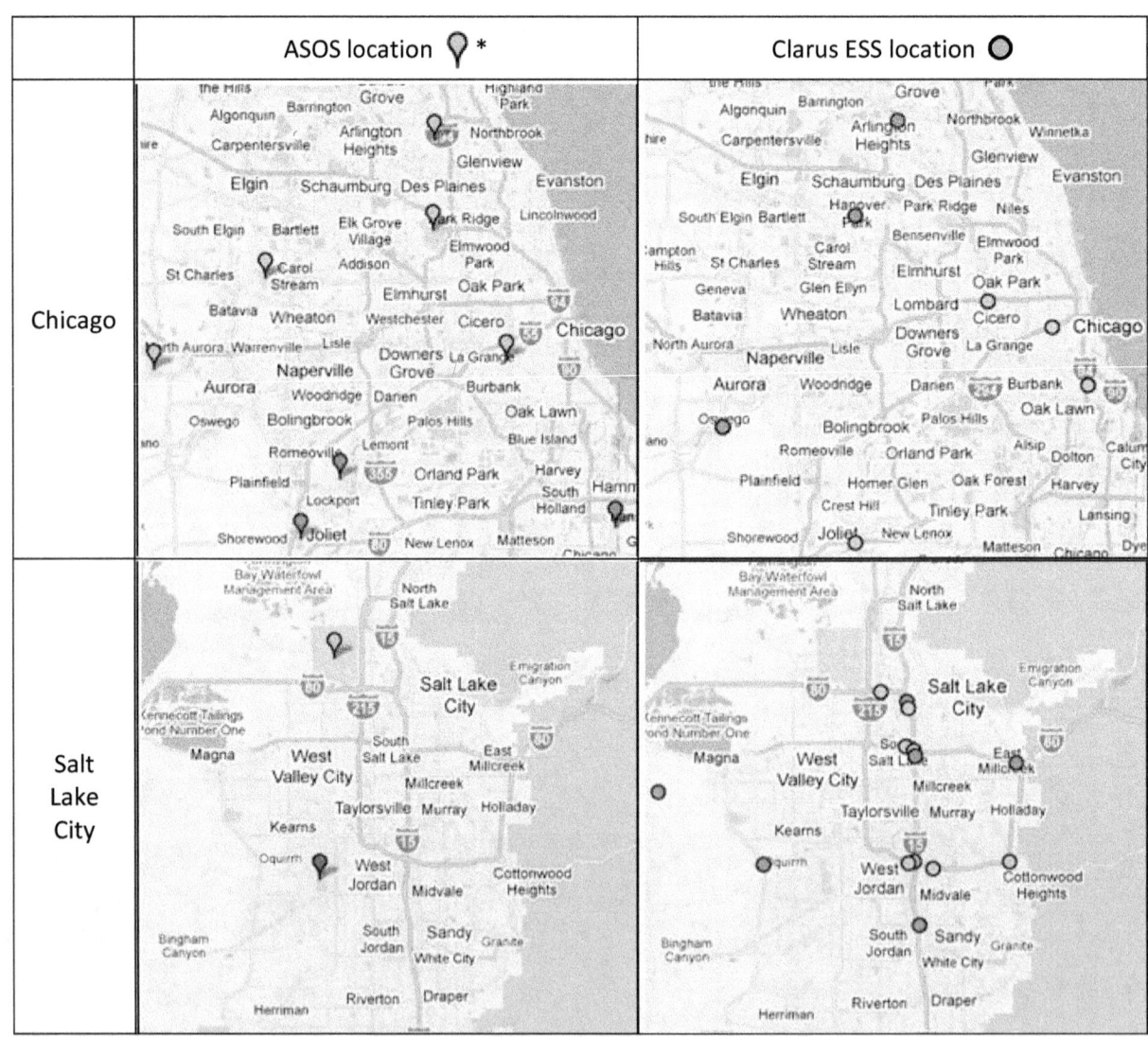

New York	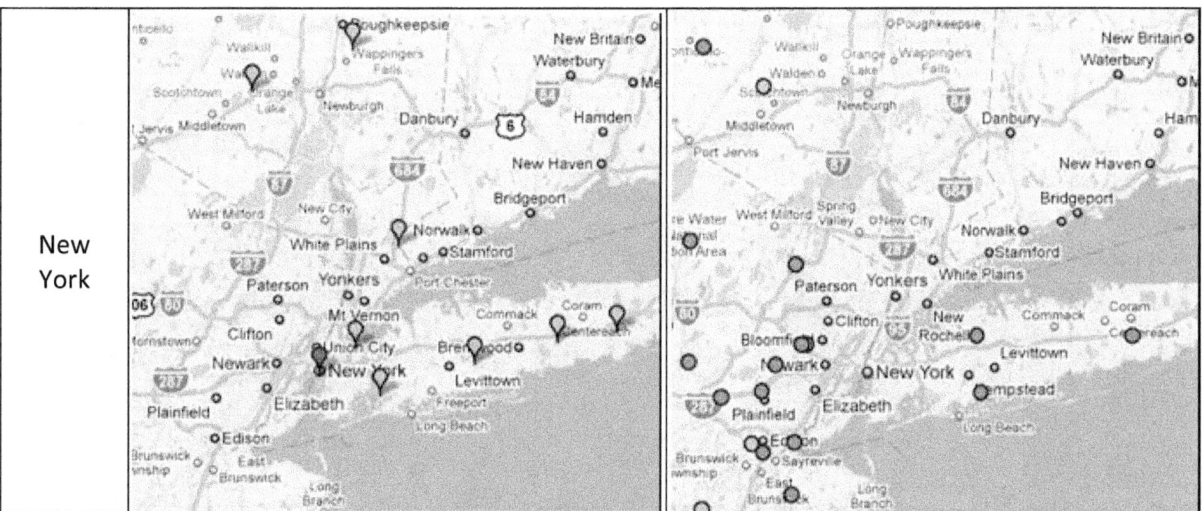	

* For the Clarus ESS location map, the grey circle indicates unavailable ESS (observed at November 18, 2010).
(Source: FAA, Surface Weather Observation Stations(leftcolumn);*Clarus* System (right column))

	ASOS location	Clarus ESS location *
Portland	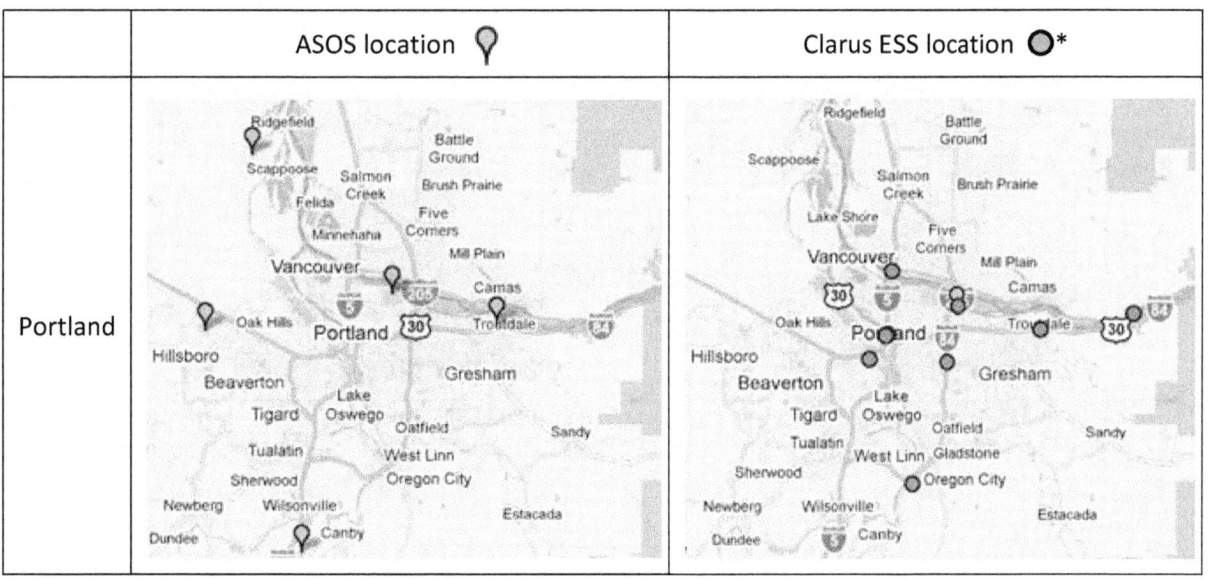	

| Appendix B | Distribution of Weather Stations (ASOS vs. Clarus) |

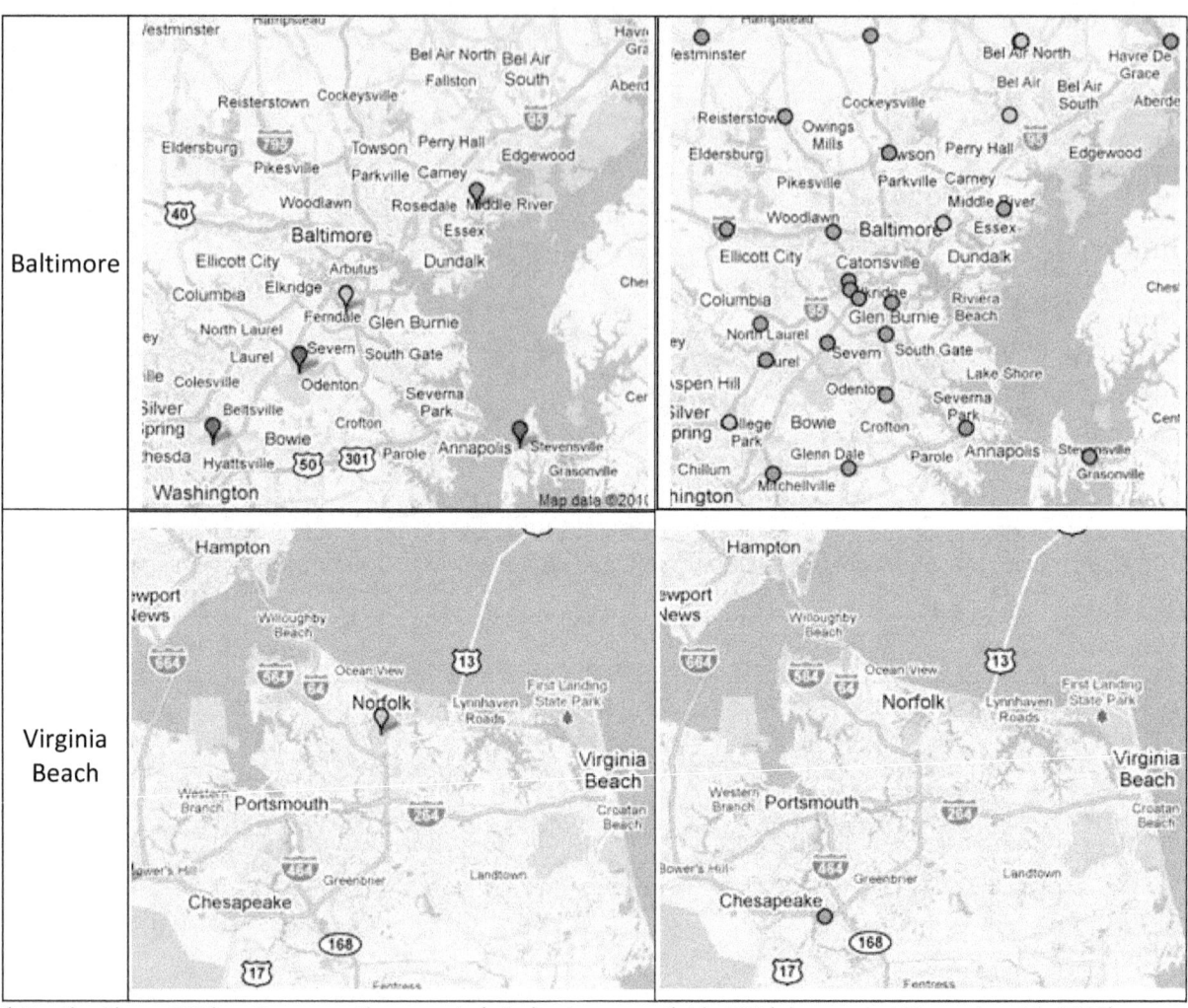

* For the Clarus ESS location map, the grey circle indicates unavailable ESS (observed at November 18, 2010).
(Source: FAA, Surface Weather Observation Stations (left column); *Clarus* System (right column))

Appendix C: Calibration Results for Traffic Flow Model and WAF

C.1 Irvine

Highway	Station ID	Weather Condition	q_{max} (veh/5-min)	v_f (mph)	alpha	k_{bp} (vpmpl)	u_f (mph)	v_0 (mph)	k_j (vpmpl)	# of observations		RMSE	R^2	WAF				
										regime 1	regime 2			F_qmax	F_vf	F_alpha	F_kbp	F_uf
I-405	1201217	normal	835	110.75	7.13	16.03	69.45	10	225	513	1775	4.81	0.90	1.00	1.00	1.00	1.00	1.00
		light rain	735	103.96	7.13	15.59	66.30	10	225	163	298	4.67	0.92	0.88	0.94	1.00	0.97	0.95
		moderate rain	647	98.15	7.13	14.89	64.07	10	225	75	46	5.68	0.84	0.77	0.89	1.00	0.93	0.92
		heavy rain	605	90.15	7.13	10.90	66.24	10	225	13	19	5.13	0.78	0.72	0.81	1.00	0.68	0.95
I-405	1201222	normal	761	128.39	8.24	19.04	67.14	10	225	1342	1166	4.17	0.84	1.00	1.00	1.00	1.00	1.00
		light rain	735	117.33	8.24	16.89	66.41	10	225	252	209	5.14	0.79	0.97	0.91	1.00	0.89	0.99
		moderate rain	758	107.04	8.24	15.10	64.74	10	225	89	32	4.12	0.79	1.00	0.83	1.00	0.79	0.96
		heavy rain	577	104.10	8.24	15.03	63.22	10	225	28	4	3.85	0.82	0.76	0.81	1.00	0.79	0.94
I-405	1201227	normal	189	70.45	5.37	5.90	62.40	10	225	470	2038	6.10	0.52	1.00	1.00	1.00	1.00	1.00
		light rain	139	66.90	5.37	3.74	62.01	10	225	133	328	4.69	0.70	0.74	0.95	1.00	0.63	0.99
		moderate rain	116	64.05	5.37	3.18	64.05	10	225	26	95	4.71	0.69	0.61	0.91	1.00	0.54	1.03
		heavy rain	107	63.64	5.37	2.01	63.64	10	225	7	25	3.69	0.70	0.57	0.90	1.00	0.34	1.02
I-405	1201229	normal	160	93.13	8.36	11.61	63.37	10	225	1787	721	4.88	0.72	1.00	1.00	1.00	1.00	1.00
		light rain	141	79.67	8.36	6.82	63.85	10	225	255	206	4.74	0.78	0.88	0.86	1.00	0.59	1.01
		moderate rain	120	73.52	8.36	4.65	63.34	10	225	81	40	4.53	0.62	0.75	0.79	1.00	0.40	1.00
		heavy rain	110	69.30	8.36	2.33	64.36	10	225	12	20	3.34	0.75	0.69	0.74	1.00	0.20	1.02
I-405	1201399	normal	871	92.52	4.57	14.61	70.73	10	225	435	1853	4.66	0.87	1.00	1.00	1.00	1.00	1.00
		light rain	799	94.17	4.57	18.54	66.83	10	225	205	246	7.51	0.85	0.92	1.02	1.00	1.27	0.94
		moderate rain	751	79.62	4.57	10.59	65.86	10	225	51	70	7.24	0.80	0.86	0.86	1.00	0.73	0.93
		heavy rain	481	73.71	4.57	6.92	65.24	10	225	10	22	6.75	0.60	0.55	0.80	1.00	0.47	0.92
I-5	1204876	normal	149	86.66	6.95	11.19	63.79	10	225	1148	1360	7.16	0.82	1.00	1.00	1.00	1.00	1.00
		light rain	125	80.11	6.95	9.08	62.66	10	225	231	230	7.13	0.70	0.84	0.92	1.00	0.81	0.98
		moderate rain	126	82.79	6.95	11.48	60.59	10	225	93	28	10.16	0.49	0.85	0.96	1.00	1.03	0.95
		heavy rain	115	81.67	6.95	11.86	59.20	10	225	26	6	10.34	0.18	0.77	0.94	1.00	1.06	0.93
I-5	1204878	normal	987	89.99	4.89	14.50	67.73	10	225	595	832	6.11	0.83	1.00	1.00	1.00	1.00	1.00
		light rain	796	82.32	4.89	10.08	67.79	10	225	93	322	7.09	0.80	0.81	0.91	1.00	0.69	0.99
		moderate rain	679	72.26	4.89	4.07	66.94	10	225	12	85	7.87	0.66	0.69	0.80	1.00	0.28	0.99
		heavy rain	620	66.78	4.89	4.07	61.93	10	225	0	23	6.33	0.48	0.63	0.74	1.00	0.28	0.91

Appendix C

Calibration Results for Traffic Flow Model and WAF

C.2 Salt Lake City

Highway	Station ID	Weather Condition	q_{max} (veh/5-min)	v_f (mph)	alpha	k_{bp} (vpmpl)	u_f (mph)	v_0 (mph)	k_j (vpmpl)	# of observations regime 1	# of observations regime 2	RMSE	R^2	WAF F qmax	WAF F vf	WAF F alpha	WAF F kbp	WAF F uf
I-15	1 - NB	normal	735	87.24	4.38	19.66	59.14	2	225	2041	381	1.86	0.88	1.00	1.00	1.00	1.00	1.00
		light rain	675	82.11	4.38	17.53	58.18	2	225	622	182	2.91	0.78	0.92	0.94	1.00	0.89	0.98
		moderate rain	690	82.84	4.38	19.04	56.90	2	225	368	20	3.09	0.37	0.94	0.95	1.00	0.97	0.96
		light snow	565	69.51	4.38	11.92	55.20	2	225	417	721	9.16	0.53	0.77	0.80	1.00	0.61	0.93
		moderate snow	514	68.08	4.38	13.37	52.54	2	225	96	62	7.41	0.49	0.70	0.78	1.00	0.68	0.89
I-15	1 - SB	normal	741	88.76	4.56	20.95	57.57	2	225	1377	776	2.73	0.63	1.00	1.00	1.00	1.00	1.00
		light rain	710	82.61	4.56	19.07	55.83	2	225	377	427	3.92	0.59	0.96	0.93	1.00	0.91	0.92
		moderate rain	633	81.22	4.56	18.91	55.10	2	225	318	70	3.45	0.50	0.85	0.92	1.00	0.90	0.91
		light snow	701	73.98	4.56	18.05	51.16	2	225	425	713	8.48	0.46	0.95	0.83	1.00	0.86	0.85
		moderate snow	588	69.35	4.56	16.75	49.34	2	225	96	62	8.67	0.11	0.79	0.78	1.00	0.80	0.82
I-215	2 - NB	normal	549	129.17	6.44	20.89	70.08	2	225	2242	174	1.99	0.80	1.00	1.00	1.00	1.00	1.00
		light rain	530	121.65	6.44	19.83	68.04	2	225	745	59	2.75	0.79	0.94	0.94	1.00	0.94	0.97
		moderate rain	555	119.33	6.44	19.91	66.58	2	225	369	19	3.44	0.62	0.98	0.92	1.00	0.95	0.95
		light snow	413	87.19	6.44	12.29	61.32	2	225	836	302	9.74	0.47	0.73	0.67	1.00	0.58	0.88
		moderate snow	397	81.91	6.44	11.71	58.64	2	225	135	23	8.95	0.11	0.70	0.63	1.00	0.56	0.84
I-215	2 - SB	normal	525	115.44	5.77	19.77	68.87	2	225	2199	218	2.85	0.55	1.00	1.00	1.00	1.00	1.00
		light rain	478	107.73	5.77	17.87	67.58	2	225	743	61	2.43	0.64	0.91	0.93	1.00	0.90	0.98
		moderate rain	420	107.03	5.77	18.45	66.12	2	225	376	12	2.97	0.60	0.80	0.92	1.00	0.93	0.96
		light snow	549	91.21	5.77	15.08	61.78	2	225	796	342	9.76	0.44	1.05	0.79	1.00	0.76	0.90
		moderate snow	407	87.41	5.77	14.63	59.95	2	225	144	14	8.47	0.42	0.78	0.76	1.00	0.74	0.87
I-80	3 - EB	normal	287	76.36	3.68	5.89	69.45	2	225	1580	837	2.14	0.41	1.00	1.00	1.00	1.00	1.00
		light rain	259	74.45	3.68	6.23	67.34	2	225	657	147	2.93	0.47	0.90	0.98	1.00	1.06	0.97
		moderate rain	226	73.64	3.68	7.40	65.35	2	225	378	10	3.94	0.21	0.79	0.96	1.00	1.26	0.94
		light snow	250	66.84	3.68	6.43	60.28	2	225	804	334	10.02	0.28	0.87	0.88	1.00	1.09	0.87
		moderate snow	152	66.38	3.68	8.91	57.50	2	225	152	6	10.56	0.05	0.53	0.87	1.00	1.51	0.83
I-215	4 - SB	normal	605	103.29	4.24	20.96	68.93	2	225	2195	124	2.06	0.47	1.00	1.00	1.00	1.00	1.00
		light rain	563	91.69	4.24	17.61	65.50	2	225	779	25	3.94	0.09	0.93	0.89	1.00	0.84	0.95
		moderate rain	477	82.54	4.24	13.40	64.10	2	225	364	24	3.69	0.12	0.79	0.80	1.00	0.64	0.93
		light snow	478	70.07	4.24	6.87	61.69	2	225	404	734	9.63	0.33	0.79	0.68	1.00	0.33	0.89
		moderate snow	380	70.44	4.24	9.50	59.01	2	225	117	41	8.72	0.14	0.63	0.68	1.00	0.45	0.86
SR-201	9 - WB	normal	557	121.93	3.94	28.49	72.35	2	225	1501	766	5.15	0.90	1.00	1.00	1.00	1.00	1.00
		light rain	555	117.61	3.94	29.72	68.16	2	225	692	112	6.34	0.81	1.00	0.96	1.00	1.04	0.94
		moderate rain	480	102.62	3.94	21.49	69.75	2	225	352	36	7.25	0.66	0.86	0.84	1.00	0.75	0.96
		light snow	433	83.81	3.94	17.59	61.36	2	225	685	453	10.99	0.36	0.78	0.69	1.00	0.62	0.85
		moderate snow	425	77.37	3.94	13.70	60.84	2	225	124	34	8.79	0.18	0.76	0.63	1.00	0.48	0.84

C.3 Chicago

Highway	Station ID	Weather Condition	q_{max} (veh/5-min)	v_f (mph)	alpha	k_{bp} (vpmpl)	u_f (mph)	v_0 (mph)	k_j (vpmpl)	# of observations		RMSE	R^2	WAF					
										regime 1	regime 2			F_qmax	F_vf	F_alpha	F_kbp	F_uf
I-94	1113	normal	591	89.15	3.92	20.88	61.48	2	225	654	1074	6.37	0.78	1.00	1.00	1.00	1.00	1.00
		light rain	579	90.10	3.92	23.51	57.11	2	225	727	1002	5.79	0.86	0.98	1.01	1.00	1.13	0.93
		moderate rain	486	78.46	3.92	21.43	52.90	2	225	78	166	4.42	0.80	0.82	0.88	1.00	1.03	0.86
		light snow	576	99.10	3.92	20.65	60.27	2	225	306	418	9.09	0.79	0.97	1.11	1.00	0.99	0.98
		moderate snow	399	78.96	3.92	23.00	52.41	2	225	5	86	13.30	0.68	0.68	0.89	1.00	1.10	0.85
I-94	1021	normal	558	110.36	5.12	19.09	70.82	2	225	490	1469	5.96	0.86	1.00	1.00	1.00	1.00	1.00
		light rain	522	102.89	5.12	18.76	65.95	2	225	252	988	5.60	0.92	0.94	0.93	1.00	0.98	0.93
		moderate rain	480	98.89	5.12	17.95	64.27	2	225	35	150	5.50	0.92	0.86	0.90	1.00	0.94	0.91
		light snow	531	110.44	5.12	21.11	62.80	2	225	385	444	7.95	0.76	0.95	1.00	1.00	1.11	0.89
		moderate snow	369	88.23	5.12	15.44	56.83	2	225	5	53	6.86	0.83	0.66	0.80	1.00	0.81	0.80
I-94	1034	normal	591	84.90	3.74	20.35	60.16	2	225	686	951	3.91	0.67	1.00	1.00	1.00	1.00	1.00
		light rain	579	85.14	3.74	19.00	58.35	2	225	365	815	4.89	0.93	0.98	1.00	1.00	0.93	0.97
		moderate rain	570	82.92	3.74	24.65	48.99	2	225	263	197	4.96	0.90	0.96	0.98	1.00	1.21	0.81
		light snow	555	89.30	3.74	21.50	55.07	2	225	315	416	5.40	0.91	0.94	1.05	1.00	1.06	0.92
		moderate snow	420	79.49	3.74	20.52	48.94	2	225	92	107	6.14	0.76	0.71	0.94	1.00	1.01	0.81
I-290	3105	normal	615	89.93	4.26	21.68	59.10	2	225	1118	631	3.83	0.40	1.00	1.00	1.00	1.00	1.00
		light rain	610	100.38	4.26	24.39	54.78	2	225	909	202	3.49	0.56	0.99	1.12	1.00	1.12	0.93
		moderate rain	530	74.74	4.26	16.95	54.13	2	225	197	186	4.30	0.79	0.86	0.83	1.00	0.78	0.92
		light snow	609	117.31	4.26	23.55	52.90	2	225	1301	174	4.45	0.32	0.99	1.30	1.00	1.09	0.90
		moderate snow	437	92.47	4.26	17.39	50.26	2	225	60	50	4.75	0.62	0.71	1.03	1.00	0.80	0.85
I-90	2030	normal	788	88.69	4.65	16.67	62.60	2	225	252	1437	3.08	0.98	1.00	1.00	1.00	1.00	1.00
		light rain	780	85.85	4.65	16.16	60.77	2	225	218	893	3.16	0.98	0.99	0.97	1.00	0.97	0.97
		moderate rain	656	79.76	4.65	11.82	61.69	2	225	9	176	2.30	0.97	0.83	0.90	1.00	0.71	0.99
		light snow	736	82.56	4.65	14.81	59.03	2	225	95	551	4.70	0.94	0.93	0.93	1.00	0.89	0.94
		moderate snow	592	91.22	4.65	24.93	46.86	2	225	6	52	2.45	0.98	0.75	1.03	1.00	1.50	0.75

C.3 Chicago (continued)

Highway	Station ID	Weather Condition	q_{max} (veh/5-)	v_f (mph)	alpha	k_{bp} (vpmpl)	u_f (mph)	v_0 (mph)	k_j (vpmpl)	# of observations regime 1	# of observations regime 2	RMSE	R^2	WAF F_qmax	F_vf	F_alpha	F_kbp	F_uf
I-90	2031	normal	812	91.63	4.66	18.95	61.50	2	225	262	1386	2.99	0.98	1.00	1.00	1.00	1.00	1.00
		light rain	748	85.14	4.66	16.08	60.75	2	225	233	910	3.18	0.98	0.92	0.93	1.00	0.85	0.99
		moderate rain	780	78.15	4.66	17.71	54.06	2	225	17	209	3.21	0.95	0.96	0.85	1.00	0.93	0.88
		light snow	692	86.38	4.66	16.82	58.06	2	225	62	540	2.89	0.98	0.85	0.94	1.00	0.89	0.94
		moderate snow	560	61.20	4.66	14.35	45.56	2	225	0	68	5.76	0.80	0.69	0.67	1.00	0.76	0.74
I-90	2113	normal	764	84.37	4.00	19.60	59.22	2	225	526	1223	3.04	0.97	1.00	1.00	1.00	1.00	1.00
		light rain	712	81.19	4.00	17.59	58.26	2	225	234	877	3.54	0.97	0.93	0.96	1.00	0.90	0.98
		moderate rain	696	75.33	4.00	10.28	62.18	2	225	5	180	3.14	0.96	0.91	0.89	1.00	0.52	1.05
		light snow	676	79.39	4.00	17.87	55.39	2	225	159	487	5.33	0.92	0.88	0.94	1.00	0.91	0.94
		moderate snow	552	75.59	4.00	19.17	49.51	2	225	2	56	3.24	0.97	0.72	0.90	1.00	0.98	0.84
I-90	2120	normal	828	86.45	3.99	20.16	60.10	2	225	525	1266	3.19	0.96	1.00	1.00	1.00	1.00	1.00
		light rain	808	83.48	3.99	18.01	58.83	2	225	255	925	3.93	0.95	0.98	0.97	1.00	0.89	0.98
		moderate rain	744	75.57	3.99	16.76	54.97	2	225	16	229	4.00	0.92	0.90	0.87	1.00	0.83	0.91
		light snow	832	88.35	3.99	21.54	55.25	2	225	283	452	5.12	0.93	1.00	1.02	1.00	1.07	0.92
		moderate snow	568	85.81	3.99	23.25	46.77	2	225	55	36	4.64	0.92	x	0.99	1.00	1.15	0.78
I-55	6120	normal	570	84.18	3.65	18.00	62.61	2	225	70	1827	3.05	0.79	1.00	1.00	1.00	1.00	1.00
		light rain	546	97.17	3.65	23.18	56.24	2	225	201	909	3.75	0.81	0.96	1.15	1.00	1.29	0.90
		moderate rain	507	68.73	3.65	19.86	50.50	2	225	11	215	3.67	0.62	0.89	0.82	1.00	1.10	0.81
		light snow	534	76.96	3.65	14.52	60.01	2	225	24	533	5.45	0.60	0.94	0.91	1.00	0.81	0.96
		moderate snow	405	68.82	3.65	16.49	49.72	2	225	5	86	5.91	0.48	0.71	0.82	1.00	0.92	0.79

Appendix C — Calibration Results for Traffic Flow Model and WAF

C.4 Baltimore

Location	Weather Condition	x (veh/5-r	vf (mph)	alpha	bp (vpmpl)	uf (mph)	v0 (mph)	kj (vpmpl)	# of observations regime 1	# of observations regime 2	RMSE	R2	F_qmax	F_vf	WAF F_alpha	F_kbp	F_uf
I-695 @ Joppa Rd	normal	584	84.70	3.12	12.18	71.51	2	225	383	362	3.89	0.94	1.00	1.00	1.00	1.00	1.00
	light rain	594	85.13	3.12	15.22	68.81	2	225	17	268	4.31	0.96	1.02	1.01	1.00	1.25	0.96
	moderate rain	514	81.41	3.12	16.07	65.01	2	225	8	74	5.17	0.93	0.88	0.96	1.00	1.32	0.91
	heavy rain	542	84.58	3.12	30.19	54.67	2	225	22	41	4.88	0.90	0.93	1.00	1.00	2.48	0.76
	light snow	356	63.98	3.12	3.52	61.00	2	225	140	51	8.08	0.17	0.61	0.76	1.00	0.29	0.85
	moderate snow	260	61.12	3.12	2.96	58.72	2	225	112	104	6.64	0.34	0.45	0.72	1.00	0.24	0.82
	heavy snow	258	60.08	3.12	0.00	60.08	2	225	2	149	5.40	0.29	0.44	0.71	1.00	0.00	0.84
I-695 @ Providence Rd	normal	550	85.15	3.46	19.04	65.34	10	225	462	244	4.47	0.79	1.00	1.00	1.00	1.00	1.00
	light rain	477	80.53	3.46	22.15	59.28	10	225	265	113	4.06	0.83	0.87	0.95	1.00	1.16	0.91
	moderate rain	429	79.78	3.46	24.66	56.70	10	225	87	9	4.39	0.08	0.78	0.94	1.00	1.29	0.87
	heavy rain	400	N/A	N/A	N/A	55.11	10	225	N/A	N/A	N/A	N/A	0.73	N/A	N/A	N/A	0.84
	light snow	557	74.95	3.46	17.20	59.33	10	225	807	181	7.37	0.36	1.01	0.88	1.00	0.90	0.91
	moderate snow	504	64.77	3.46	16.04	52.41	10	225	278	117	8.21	0.36	0.92	0.76	1.00	0.84	0.80
	heavy snow	332	50.77	3.46	2.31	49.34	10	225	22	142	6.06	0.72	0.60	0.60	1.00	0.12	0.76
I-695 @ Stevenson Rd	normal	676	85.34	4.81	12.84	66.80	10	225	743	265	5.52	0.58	1.00	1.00	1.00	1.00	1.00
	light rain	653	84.79	4.81	13.79	65.18	10	225	163	455	4.06	0.94	0.97	0.99	1.00	1.07	0.98
	moderate rain	559	81.20	4.81	15.54	60.47	10	225	21	48	3.37	0.93	0.83	0.95	1.00	1.21	0.91
	heavy rain	589	81.67	4.81	16.72	59.45	10	225	77	23	3.94	0.90	0.87	0.96	1.00	1.30	0.89
	light snow	608	82.86	4.81	15.16	62.10	10	225	209	65	5.54	0.57	0.90	0.97	1.00	1.18	0.93
	moderate snow	489	68.50	4.81	12.68	54.27	10	225	389	62	7.43	0.16	0.72	0.80	1.00	0.99	0.81
	heavy snow	425	65.96	4.81	13.43	51.63	10	225	125	39	6.22	0.27	0.63	0.77	1.00	1.05	0.77
I-695 between Stevenson Rd and Greenspring Ave	normal	609	79.11	3.73	10.65	67.68	10	225	423	446	5.20	0.66	1.00	1.00	1.00	1.00	1.00
	light rain	570	76.76	3.73	10.38	65.98	10	225	100	332	3.72	0.93	0.94	0.97	1.00	0.97	0.97
	moderate rain	526	74.93	3.73	10.30	64.51	10	225	16	60	3.01	0.94	0.86	0.95	1.00	0.97	0.95
	heavy rain	515	72.88	3.73	12.37	60.92	10	225	73	27	3.86	0.86	0.85	0.92	1.00	1.16	0.90
	light snow	369	65.90	3.73	6.06	60.49	10	225	72	43	5.38	0.10	0.61	0.83	1.00	0.57	0.89
	moderate snow	366	66.67	3.73	7.73	59.74	10	225	329	57	7.32	0.04	0.60	0.84	1.00	0.73	0.88
	heavy snow	394	66.66	3.73	10.87	57.11	10	225	133	27	8.94	0.04	0.65	0.84	1.00	1.02	0.84
I-695 @ US 1 Outer Loop	normal	471	88.22	3.92	15.36	67.34	2	225	269	157	3.52	0.86	1.00	1.00	1.00	1.00	1.00
	light rain	484	84.29	3.92	18.36	60.93	2	225	174	242	4.78	0.95	1.03	0.96	1.00	1.20	0.90
	moderate rain	379	71.71	3.92	7.61	62.91	2	225	7	58	5.01	0.95	0.80	0.81	1.00	0.50	0.93
	heavy rain	306	69.59	3.92	13.86	54.67	2	225	81	19	5.34	0.83	0.65	0.79	1.00	0.90	0.81
	light snow	393	81.59	3.92	15.22	62.48	2	225	122	65	7.68	0.83	0.83	0.92	1.00	0.99	0.93
	moderate snow	195	N/A	N/A	N/A	50.70	N/A	N/A	N/A	N/A	N/A	N/A	0.41	N/A	N/A	N/A	0.75
	heavy snow	279	N/A	N/A	N/A	49.72	N/A	N/A	N/A	N/A	N/A	N/A	0.59	N/A	N/A	N/A	0.74

C.4 Baltimore (continued)

Location	Weather Condition	vx (veh/5-r	vf (mph)	alpha	bp (vpmpl)	uf (mph)	v0 (mph)	kj (vpmpl)	# of observations regime 1	# of observations regime 2	RMSE	R2	F_qmax	F_vf	WAF F_alpha	F_kbp	F_uf
I-695 @ I-70	normal	543	83.11	4.53	16.36	61.95	10	225	230	174	3.88	0.91	1.00	1.00	1.00	1.00	1.00
	light rain	585	84.27	4.53	21.29	57.36	10	225	350	244	3.31	0.95	1.08	1.01	1.00	1.30	0.93
	moderate rain	514	83.40	4.53	22.70	55.35	10	225	58	18	3.01	0.89	0.95	1.00	1.00	1.39	0.89
	heavy rain	488	81.87	4.53	22.26	54.85	10	225	88	12	3.92	0.79	0.90	0.99	1.00	1.36	0.89
	light snow	521	88.28	4.53	19.51	61.92	10	225	81	33	4.06	0.86	0.96	1.06	1.00	1.19	1.00
	moderate snow	494	81.75	4.53	26.31	50.87	10	225	163	5	7.22	0.04	0.91	0.98	1.00	1.61	0.82
	heavy snow	360	54.12	4.53	8.58	47.00	10	225	37	17	6.43	0.19	0.66	0.65	1.00	0.52	0.76
I-695 approaching US 40 W	normal	650	96.75	5.00	16.81	68.83	10	225	407	140	3.61	0.80	1.00	1.00	1.00	1.00	1.00
	light rain	617	90.57	5.00	19.18	61.61	10	225	72	344	3.69	0.95	0.95	0.94	1.00	1.14	0.90
	moderate rain	560	85.45	5.00	18.63	58.97	10	225	52	58	4.41	0.93	0.86	0.88	1.00	1.11	0.86
	heavy rain	578	82.88	5.00	20.59	55.10	10	225	70	30	3.15	0.96	0.89	0.86	1.00	1.23	0.80
	light snow	538	90.69	5.00	17.29	64.10	10	225	124	82	7.54	0.56	0.83	0.94	1.00	1.03	0.93
	moderate snow	402	89.70	5.00	23.02	56.46	10	225	222	22	6.77	0.40	0.62	0.93	1.00	1.37	0.82
	heavy snow	191	N/A	N/A	N/A	43.94	N/A	N/A	N/A	N/A	N/A	N/A	0.29	N/A	N/A	N/A	0.64